BREAKDOWN PRESS
1 Berwick Street
London W1F 0DR
www.breakdownpress.com

ISBN 978-1-91108-121-0
10 9 8 7 6 5 4 3 2 1

The New Dinosaurs: An Alternative Evolution was originally published in 1988 in the UK by Eddison/Sadd Editions Limited and in the USA by Salem House Publishers (1988).

Editor: Tom Oldham
Copy Editors: Rachel Cloyne, Fraser Geesin, Douglas Noble
Designed by David Fordham
Facsimile production and additional layout by Joe Hales and Sam Eccles, Joe Hales studio
Printed in Belgium by Drukkerij Graphius

breakdownpress

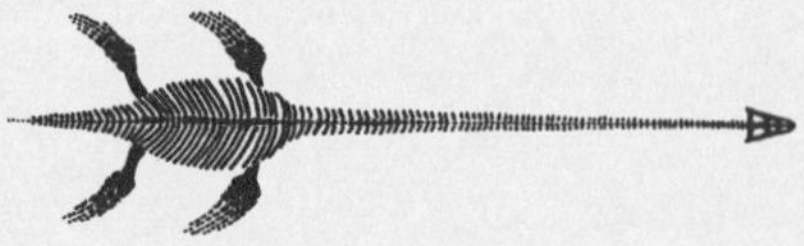

THE NEW DINOSAURS

AN ALTERNATIVE EVOLUTION

DOUGAL DIXON

FOREWORD BY DESMOND MORRIS

breakdownpress

FOR
LINDSAY

AND
MARY

CONTENTS

AUTHOR'S INTRODUCTION

IN 1988 an expedition was launched, into an alternative universe in which the meteorite missed, the great extinction that marked the end of the Cretaceous period and the beginning of the Cenozoic never took place, the Age of Reptiles did not come to an end and the Age of Mammals never began. The dinosaurs, or in modern parlance, the *non-avian* dinosaurs, as well as all the other inhabitants of the time, large and small, continued to thrive and evolve until the present day.

The report of this expedition was published at the time. Analysis of the observations was based on paleontological knowledge that was then available to the researchers. In the three and a half decades since the report was released, dinosaur science has moved on. Whole new families of dinosaurs have been recognized. Many of the conclusions of the report have been challenged in the light of subsequent research.

In recent years another expedition has been mounted to this same alternative universe, and the fauna – the same fauna – seen there was studied with the added scientific knowledge that has been amassed since 1988.

The geographic and biological background – the biogeography – of the original report is still sound. The continental movements, the climatic zones and the distribution of wildlife, along with their patterns of migrations and dispersal, were adequately interpreted. Indeed these factors are the same as we find on Earth in our own universe. It is the descriptions of the animal life that have had to be revised.

Now it is time for the revised report to be published.

– Dougal Dixon, 2024

This is a facsimile reproduction of the 1988 first edition of The New Dinosaurs by Dougal Dixon. Some of the text has been changed at the request of the author to reflect scientific discoveries made in the intervening years. The changes have been made in a slightly different typeface to make their presence clear.

FOREWORD

For the many who enjoyed, as I did, Dougal Dixon's first book of imaginary animals – AFTER MAN, A ZOOLOGY OF THE FUTURE – his new volume will provide further delights. There is the same technical skill, the same brilliant inventiveness and, without question, the same joyous celebration of the evolutionary process. As before, the author has never allowed his imaginative leaps to go beyond the bounds of biological possibility, and it is this self-discipline that puts his fictitious fauna streets ahead of anything found elsewhere, in either science fiction writing or in Hollywood films about 'lost worlds' or primeval struggles.

As a zoologist I nearly always find myself disappointed by other people's imaginary animals. The monsters they concoct are usually boringly obvious or totally improbable. But Dougal Dixon knows exactly what he is doing and gives such conviction to his creatures that it is sometimes hard to convince oneself that, somewhere, at some time or other, they have not really existed – and perhaps even exist today, if only we could be lucky enough to discover them.

With each turn of the page I am excited to find what new form will greet my eyes, and I am rarely disappointed. The only let-down is the realization that, despite their amazingly detailed representations, we will never be able to come face to face with a living Dougaloid. But no matter, they will always be here between the covers of his books, for us to enjoy whenever we feel the need to release ourselves from the tyranny of the world as it is, and enjoy how it might have been.

By making a single, simple alteration in the earth's prehistory, namely the absence of the catastrophe that wiped out the great dinosaurs, Dixon has set us off on a new evolutionary voyage – a truly creative endeavour and one which I have relished from start to finish. What he has done – with both AFTER MAN and THE NEW DINOSAURS is so exciting that I feel it deserves a special name of its own: Alternative Zoology. And I for one cannot wait for the next volume to appear, when, who knows, the author may take us off to another planet altogether, where a parallel evolutionary process has taken place, guided by the special environmental conditions that exist there. In Dixon's safe hands, what a wonderful safari that could be.

Desmond Morris

The Great Extinction

The Theories

The dinosaurs were some of the most magnificent creatures that ever lived. They evolved in the late Triassic period, about 220 million years ago, and from humble crocodile-like origins, became the most varied and abundant animals on Earth.

Nimble, darting, brightly coloured meat-eaters chased lizards through the ferny undergrowth. Big meat-eaters strode dragon-like through the forests hunting slow-moving prey. Huge long-necked plant-eaters, veritable mountains of flesh, roamed open plains and woodlands in family groups, browsing from the tops of the trees, and keeping wary eyes open for the predacious hunters. Small, fleet-footed plant-eaters sprinted on their long hind legs from one patch of vegetation to another, snatching at leaves and shoots, and quickly ducking away when danger threatened. Lumbering armoured plant-eaters, safe behind their flamboyant and colourful plates and horns, chomped and chewed the prolific vegetation in the warm and equable climates of the long, tranquil Jurassic and Cretaceous periods.

For over 180 million years they were the most successful life form on the planet.

And then they all disappeared.

The Great Extinction, as their disappearance has come to be known, befell the dinosaurs at the end of the Cretaceous period, about 65 million years ago. At this time not just the dinosaurs but about 75 per cent of all living creatures were wiped from the face of the Earth.

The rocks supply the evidence. The strata that date from before this time contain the fossils of the kinds of animals that had been present for 150 million years previously. Rocks that formed from continental deposits, such as shales and mudstones from river sediments, and sandstones from deserts, contain the remains of dinosaurs and flying pterosaurs. Rocks that formed in the sea, such as limestones and chalk, contain the fossils of sea reptiles like plesiosaurs and mosasaurs, the ubiquitous pterosaurs, and invertebrates like the tentacled shellfish, the ammonites.

Then there is a break in the strata and the rocks immediately above exhibit a different picture of life on Earth. No dinosaurs or pterosaurs are evident here. Nor are there fossils of marine reptiles or ammonites. Something had happened to change things completely. This break gives a useful boundary for geological dating. The time before the break is called the Cretaceous period of the Mesozoic era. *Creta* is the Greek word for chalk and the period is named for the deposits of chalk that were laid down in the sea at this time. The time immediately after the break is called the Palaeocene epoch of the Cenozoic period. Palaeocene means the ancient epoch of modern life and the 'modern life' referred to here means, basically, the mammals.

Up to this point the mammals had been very small and insignificant, scuttling mouse-like among the feet of the dinosaurs and scrambling squirrel-like up trees beyond their reach. Throughout the 150 million-year Age of Reptiles they did not amount to much at all. Then, as the dinosaurs and the other great reptiles disappeared, they came into their own. It is the way of evolution that, once a particular animal dies out, something else will soon develop to take its place. Wherever there is a food source that is not exploited, an animal will evolve to exploit it. With the total disappearance of the great reptiles, there was a complete replacement by new creatures, and these new creatures were the mammals. Pigs and elephants evolved to take the place of the herbivorous dinosaurs. Strange wolf-like mammals called creodonts evolved to prey on them, taking the place of the carnivorous dinosaurs. In the absence of pterosaurs, the bats evolved. In the seas, whales and seals developed to take the place of the plesiosaurs and mosasaurs. The Age of Reptiles had clearly given way to the Age of Mammals.

How had this great change come about? It was certainly nothing to do with the mammals becoming stronger and more successful and hence ousting the reptiles. As we have seen, it was the other way round. Something else had killed off the creatures that had been established for 150 million years. The cause may have been sudden, or it may have been gradual. The geological record is deceptive when it comes to calculating time. A million years is a mere eye-blink and may be represented by a bed of rock a few centimetres thick – or even no rock at all.

The generally accepted theory now is that, 66 million years ago, a meteorite hurtled in from deep space and struck the Earth in the region that is now the Yucatan peninsula in Mexico. At that time the area consisted of tropical islands in a shallow gentle shelf sea. The meteorite was about 10 kilometres in diameter and came in at an angle of between 20 and 30 degrees from the south-east.

The initial blast of heat generated by the impact formed a fireball that was in excess of 10,000 degrees, and seared away all nearby life instantaneously. Solid rock immediately liquified, sulphur-rich sediments of the continental shelf were vapourised, the underlying granites of the upper crust melted and ran, hollowing out a crater 100 kilometres in diameter and bringing up molten material from 14 kilometres down.

The shock wave spread out in the form of winds ripping through at 1000 kilometres per hour, shredding vegetation and scouring soils up to 1500 kilometres away across the nearby continents. The shock circled the globe, reaching New Mexico 11 minutes later, and southern Patagonia after 42 minutes. Then, as the fireball arose as a mushroom cloud, the winds rushed back into the vacuum below with a similarly devastating blast.

The sea was swept up into a series of tsunamis, anything between 50 and 300 metres high that crashed into nearby coastlines and penetrated more than 100 kilometres inland, carrying fragments of torn-up seabed into the depths of the continents. Tsunamis were generated all around the world within two hours, due to the seiche effect of water-transmitted shock waves in narrow seaways.

Then the debris started falling. Incandescent lumps, molten and solid, rained down. Because of the angle and the direction of the impact, the continent of North America carried the brunt of the debris that was blasted out. For three days hot rocky material was swept around the Earth in red-hot clouds, heating the atmosphere, dropping incendiary particles and drying out and igniting the vegetation beneath, causing world-wide eruption of forest fires.

Lighter material stayed aloft and choked the atmosphere for perhaps a year, shutting out the sunlight, darkening the land and killing off any surviving vegetation. Sulphur chemicals in the dust derived from the pulverised seabed were then washed out by water as vapour condensed, and fell as acid rain. Heavy metals settled out, poisoning what was left of the soil. With the sunlight blocked and reflected away by the clouds the climate cooled and the dust-choked land became dark and bitterly cold. Much later the carbon dioxide, methane, water vapour and other gases released by the impact and remaining in the atmosphere produced an intense greenhouse effect, making the air hot and stuffy once more.

The chemicals destroyed the ozone layer – that portion of the atmosphere that lies in the lower part of the stratosphere between about 15 and 35 kilometres above the Earth's surface – and is comparatively rich in ozone, a kind of oxygen. Normally this layer absorbs over 90 percent of the lethal ultra-violet radiation from the sun. Unblocked, ultraviolet radiation penetrates deep into living bodies and disrupts DNA molecules in functioning cells, giving rise to increased susceptibility to cancers. It also destroys the delicate mechanisms that are used for the sense of sight in most animals. In short, the ozone layer keeps the Earth's surface suitable for life, and once destroyed it would be years before the layer could regenerate.

Meanwhile the Earth's crust shook from the impact, and the disruption spread as earthquakes around the world, and may have had a triggering effect on the widespread volcanic eruptions in the Deccan region of India – almost exactly on the other side of the globe at that time. These basaltic eruptions continued for millennia, covering half of the Indian sub-continent in plateau lavas, and their constantly

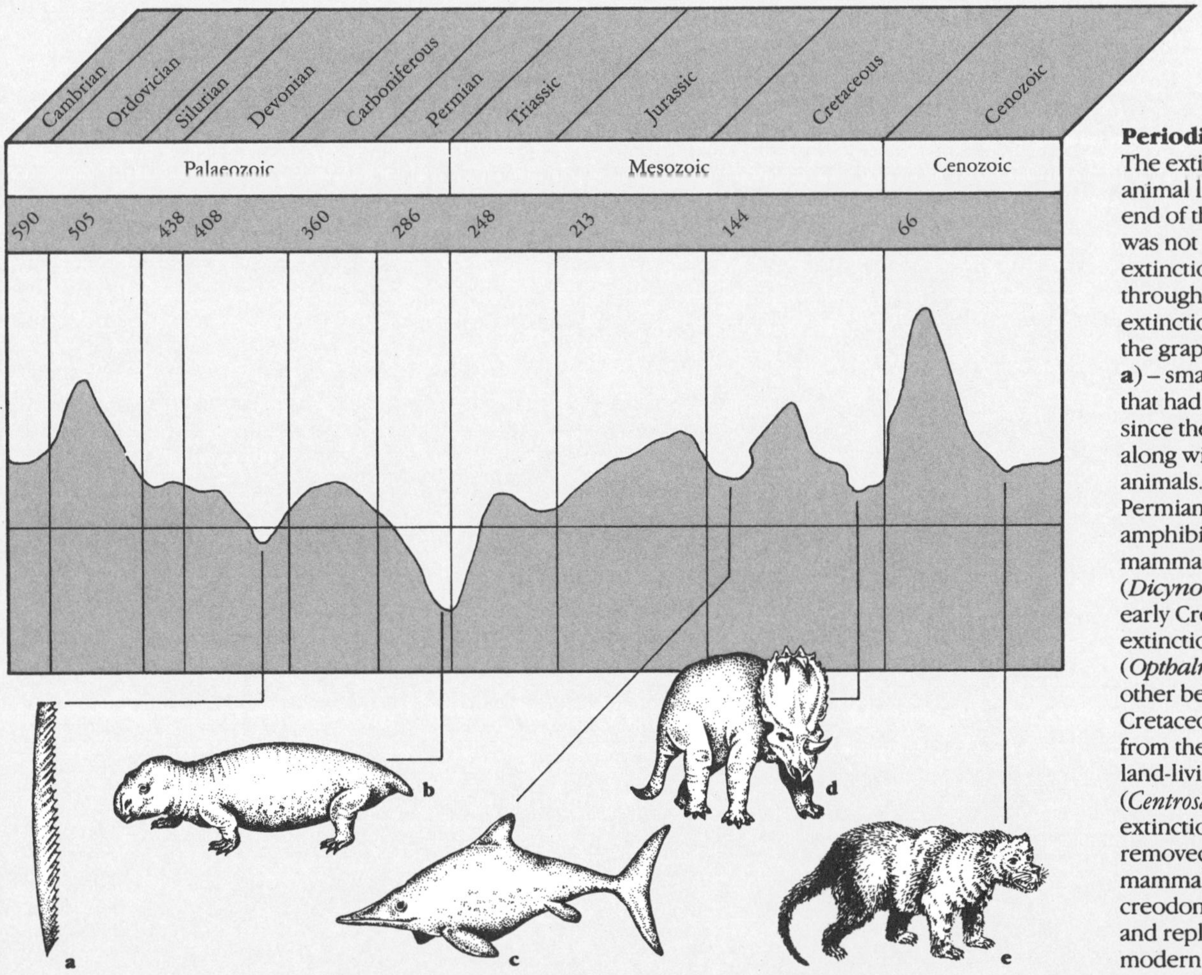

Periodic extinctions
The extinction of much of the animal life of the world at the end of the Cretaceous period was not an isolated event. Mass extinctions have occurred throughout time. In a mass extinction in Devonian times, the graptolites (*Monograptus*, **a**) – small colonial creatures that had drifted in the oceans since the Cambrian – died out along with many other sea animals. At the end of the Permian, huge land-living amphibians and most of the mammal-like reptiles (*Dicynodon*, **b**) died out. The early Cretaceous saw the extinction of ichthyosaurs (*Opthalmosaurus*, **c**) among other beasts. The late Cretaceous event was important from the point of view of large land-living animals (*Centrosaurus*, **d**), but a further extinction in the Tertiary removed many primitive mammals, like the meat-eating creodonts (*Sarkastodon*, **e**), and replaced them with more modern types.

generated gases – greenhouse gases such as carbon dioxide and water vapour as well as poisonous substances like hydrogen sulphide and sulphur dioxide – having a continuing toxic effect on the composition of the Earth's atmosphere.

Not much could have survived all this.

The plant life burned, or shrivelled, or perished as their photosynthetic metabolism collapsed through lack of sunlight. Plant-eating animals had nothing to eat, and so those that survived the initial devastation starved to death. Meat-eating animals subsequently had nothing to eat either, and they died. It was the end of the big animals, the end of the dinosaurs.

In the seas the phytoplankton – the microscopic floating plant life – was severely restricted by the lack of sunlight. The acid rain was so intense that the ocean surface became highly acidic and many of the shell-forming animals could not form shells. Such small plants and animals were the base of the food chains here, and so the medium-sized animals that fed on them, and subsequently the big animals that fed on these, could not survive. Many animals, such as the ammonites, relied on surface currents to spread their eggs, and this part of their lifecycle was lost. The Cretaceous creatures of the oceans disappeared.

What happened then?

Not everything died. Smaller animals survived, possibly protected in burrows and feeding on long-lasting seeds and plant debris. Forests were all gone, burned away by the planet-wide bush fires. From these surviving creatures evolved the animal life to come, and most of these were mammals. The replacement vegetation consisted mostly of ferns – it would be a million years before forests were re-established. The conifer-dominated tropical forest of the Amazon basin were wiped out, and eventually replaced by the kind of flowering plants that we find there today. Perching birds that thrived at the end of the Cretaceous period, were rendered extinct by the loss of trees. The birds that survived were ground dwellers, comparable to modern pheasants and chickens. The tree-living perching habit had to evolve all over again. Freshwater habitats seemed to have given some degree of protection, and most families of freshwater turtles and crocodiles survived the disaster. Within a few million years nature had reset itself and the ecosystems thrived once more.

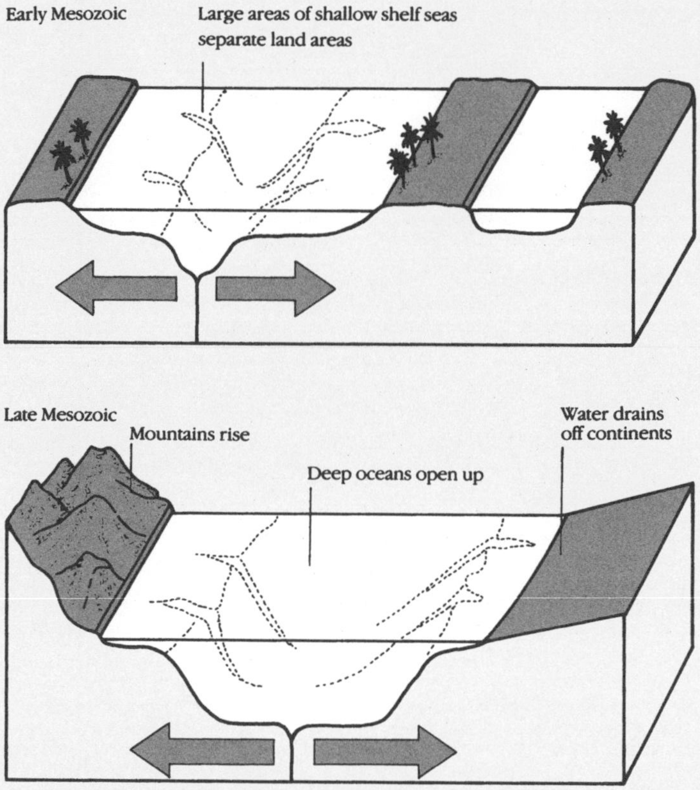

The influence of continental drift
During the early Mesozoic, the continents were close to one another and much of their area was covered with shallow sea. The warm surface waters supported an abundance of animal and plant life. In the late Mesozoic, the continents were drifting apart and deep sea areas developed between them. The shallow shelf seas drained away into the colder deep sea, and mountains rose up. With the loss of the surface water habitat, sea plants and animals died.

How do we know all this?

The first evidence came to light in the 1980s when it was found that the topmost Cretaceous beds around the world all contained a very high proportion of the metal iridium. Iridium is rare at the Earth's surface but is common in meteorites. No matter what the rock was, or how it was deposited, as long as it dated from this time it contained the iridium.

Soon after it was found that this layer also contained tiny discs and spheres of glass called tektites, formed when hot molten material solidifies in flight. In this instance they were formed by the instantaneous melting and dispersal of the Earth's crust at the impact site. There was also a widespread occurrence of quartz crystals that showed the kinds of fractures that would be expected if such crystals had been subjected to extreme shock. The implication was that a great iridium-rich meteorite had disintegrated in an event that had spread shocked and molten material around the world.

In 1990 the site of the impact was found. This was a buried crater, 200 kilometres in diameter, on the coast of the Yucatan Peninsula in Mexico. With no visible sign on the surface, it was found by geophysical investigation. The timing of its formation was right. The size conformed to what was predicted from the world-wide deposits. Its slightly keyhole shape indicated the direction from which it had come, and the angle at which it impacted.

Since then there has been a torrent of evidence that has backed up the theory. The world-wide eruption of wildfires at this time was indicated by the study of contemporary soot and carbon from New Zealand to Colorado. The studies took place between 2000 and 2013 and the isotopes from this carbon matched those from carbon deposits from forest fires. In 2014 isotope studies of marine sediments in Texas indicate a sudden cooling of the sea in the months or decades immediately following the impact.

Widespread occurrences of gypsum across the South American continent dating from the time, were the result of the sulphur-rich particles in the atmosphere, derived from the shallow sea deposits at the site of the impact. Drilling at the site shows that all the sulphur-rich sediments were blasted away, and that the granite of the basement was melted and resolidified.

Evidence for tsunami damage was reported from a geological excavation in North Dakota in 2019. The sequence of rocks a metre and a half thick formed a bed that represented the mouth of a river at

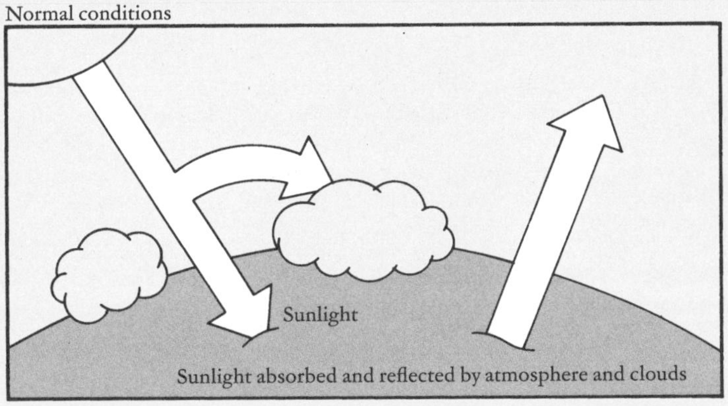

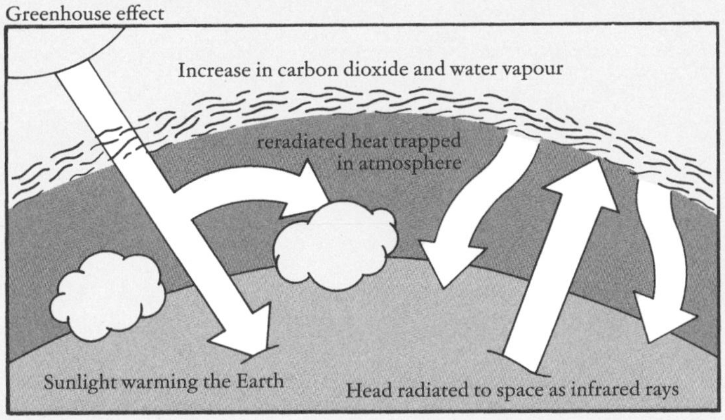

Greenhouse effect
The different proportions of gases found in the atmosphere can have a profound effect on the climate. Normally, temperatures are kept in balance. Of the sunlight that reaches the Earth's surface, some is reradiated into space as infrared rays. An increase in the proportion of atmospheric carbon dioxide or water vapour causes much of the reradiated infrared heat to be absorbed and trapped in the atmosphere, and hence the Earth's surface heats up.

the very end of the Cretaceous period. It showed, first of all, a rain of molten rock fragments so heavy that it ignited the local vegetation, then a 10-metre-high wave that swept freshwater fish upstream and left them stranded on a sand bar. This was followed by another great wave that buried everything in marine sediments complete with the remains of sea-living ammonites and reptiles, all happening within a time frame of minutes or hours. And all this time the hot molten material was still raining down.

The subsea earthquake that generated the tsunami in 2011 that caused such damage in Japan produced 1.5 metre high seiche waves in Norwegian fjords half an hour later – the other side of the world and from a shock that would have been minuscule compared with the impact of the Yucatan meteorite.

Other ideas

Although the meteorite impact is accepted by nearly all scientific bodies, there were always other theories as to what killed the dinosaurs. The alternative theory that involves spectacular disaster is suggested by the presence of the Deccan Plateau lavas in India. These erupted continuously for 30,000 years and now cover an area of 500,000 square kilometres to a depth of 2,000 metres. Before erosion they would have covered an area of 600,000 square kilometres. Their timing, at the very end of the Cretaceous Period, matches the extinction of the dinosaurs. Such intensive eruptions would have produced enough volcanic gases to alter the composition of the atmosphere and bring about a climate change that would have been fatal to many animals. However, it has been estimated that these eruptions would have reduced sunlight by about five percent, well short of the estimated 20 percent necessary to trigger the massive extinctions of the time. The eruption of similar large-scale plateau lavas in Russia between the Permian and Triassic periods have been implicated in the mass-extinction that occurred then – and these did not coincide with any meteor strike.

Les dramatic theories implicate such factors and climate change or plate tectonics. During the age of dinosaurs, the geography of the Earth changed considerably. At the very beginning, in the Triassic Period, all the landmasses were in a single continent we call Pangea. Then in the Jurassic Period this supercontinent began to break up. In the Cretaceous Period many separate continents existed and they were moving away from one another.

One result of this would be the increase in shallow continental seas. The water displaced by earth movements would have spilled over the continental edges and spread out as shelf seas. This would have had an influence on world climates, bringing cooler and moister conditions to more places, and may have had a fatal effect on big animals that had evolved to live in stable conditions over millions of years. Another result would be that moving continents would have collided with one another, exchanging faunas. New animals could have brought diseases – diseases to which they had grown immune – and spread them to the indigenous wildlife.

Added to these were statistics that seemed to show that the dinosaur population was in decline towards the end of the Cretaceous period, and that any meteorite impact would only have given the dying blow to a system already on its way out.

Then there was the idea that the suddenness of the loss of the dinosaurs was more apparent than real. Fewer sedimentary rocks were being formed at the end of the Cretaceous and so fewer dinosaurs were being fossilized. The passage of a long period of time was preserved as a very thin sequence of rocks, and so any change would have appeared more swift than it actually was.

Science used to be alive with arguments about the cause of the mass extinction at the end of the Cretaceous period. However, most of that seems to have settled, and the vast majority scientists accept the idea that the big dinosaurs disappeared in a cataclysmic meteorite impact.

Alternatively ...

For the purposes of this book none of that happened. The meteorite missed. The comet swarm passed by. The sea plants were able to adapt to the changing water temperatures. The land animals were able to resist the diseases and parasites of their neighbours.

In short, the animal life that had developed throughout the 150 million years of the Mesozoic era continued to evolve for at least another 66 million years without a break. And that is the basis for our book.

However, before we look at the creatures that exist in our alternative zoology of modern times, we must first look at what kind of animal the dinosaur was, and how it evolved and developed during its heyday, in the distant times of the Mesozoic era.

WHAT IS A DINOSAUR?

EVOLUTION OF THE LAND-LIVING REPTILES

The first vertebrate animals that could live completely on land appeared in the Carboniferous period. They laid waterproof eggs and this meant that they did not need to return to the water to breed, as their ancestors the amphibians did.

Immediately several evolutionary lines of land-living vertebrates developed which in turn, became dominant, and faded away. The first of these were the paramammals. They started off as lizard-like animals, some of which carried sails on their backs, and had both meat-eating and plant-eating forms. Later they developed a warm-blooded metabolism and became more and more like true mammals, with furry coats and the ability to bear their young alive. They used to be called the "mammal-like reptiles" but now we regard them as too different from the main reptile group to merit the name. Finally, the mammals themselves evolved out of this group.

The paramammals faded away in Triassic times, leaving only their mammal descendants. Another group, the archosaurs also appeared in the Carboniferous and existed throughout the paramammal heyday, but as insignificant lizard-like animals. With the paramammals gone these archosaurs then expanded to take their places. Strongest of these was the group that consisted of the crocodiles and their immediate relatives. Again meat-eaters and plant-eaters evolved in this group. There were lion-sized hunting crocodile relatives, armoured pig-like vegetarian crocodile relatives, scampering two-footed insectivorous crocodile relatives... and these also became warm-blooded. There were also semi-aquatic ambush hunters that reverted to a cold-blooded

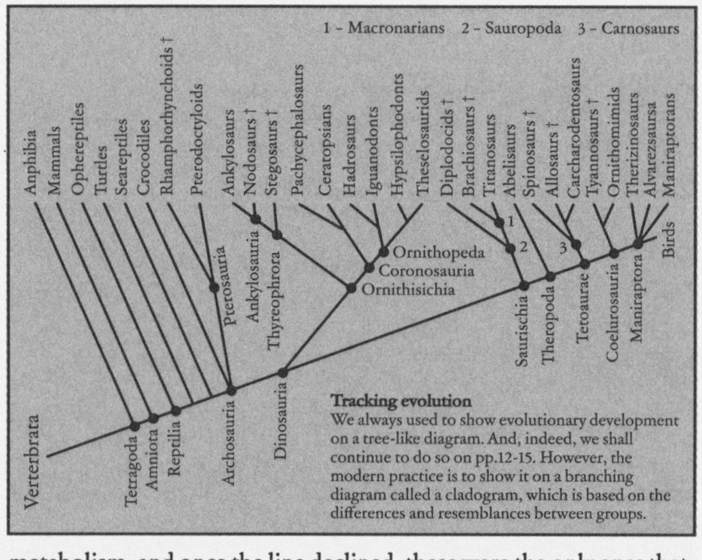

Tracking evolution
We always used to show evolutionary development on a tree-like diagram. And, indeed, we shall continue to do so on pp.12-15. However, the modern practice is to show it on a branching diagram called a cladogram, which is based on the differences and resemblances between groups.

metabolism, and once the line declined, these were the only ones that survived. They exist at the present day – as the crocodiles themselves.

Meanwhile the two-footed insectivorous archosaur group branched away and developed into two distinct lines – one that was to evolve into the pterosaurs, and the other that was to evolve into the dinosaurs. This they both did on the extinction of the dominant crocodile-relatives. And this is where the evolution of the dinosaurs really takes off.

First up were the Saurischia – the "lizard hipped" – named because of the lizard-like arrangement of the hip bones. These animals differed little from their ancestors, but the way the legs were attached to the hips, allowed the weight of the body to be carried at the top of the legs rather than slung between them as in other reptiles. This stronger arrangement allowed heavy-bodied animals to evolve and soon the standard shape was of a bipedal animal with its meat-eating jaws out the front, small grasping arms, a horizontal back all balanced by a heavy tail. This is the shape of the meat-eating dinosaurs – the theropods.

But early in their history some of these developed a plant-based diet. A plant-eating animal needs a much more sophisticated digestive system than a meat-eating animal, and so a much bigger volume of stomach and intestine. The lizard-like arrangement of the hip bones meant that the big digestive system of the vegetarian saurischian had to be carried well forward in the body. It could no longer balance on is hind legs and so it evolved a four-footed stance. At the same time the neck became long, and that allowed the animal to reach as much food as it needed. This shape – the massive four-footed long-necked plant-eating saurischian – became the second major dinosaur group – the sauropods.

Back to the beginning of the dinosaurs, a plant-eating group arose that we call the Ornithischia – the bird hipped" – named because the arrangement of hip bones resembled those of a bird. The smallest of these could remain bipedal, as the different hipbone arrangement meant that the big plant-eating gut could be carried beneath the hips rather than in front, and remain balanced by the tail.

Two main groups of the Ornithischia are currently recognized by science. The first, the cerapoda, consists of the two-footed forms, and also those with armour on the head. The second, the thyreophora, are those with armour or plates on the back.

So complex is the classification of dinosaurs, that it keeps changing according to new evidence that is discovered. Late in their history the theropods developed still further. As part of their warm-blooded metabolism they carried an insulating covering of feathery structures. In the late Jurassic and the Cretaceous periods some of the smaller species developed these into display structures and spectacular feathers appeared. Some of the most lightweight species took to the trees and the feathers began to develop into structures that would allow their owners to glide from branch to branch, and even to fly with a muscular flapping motion. From these evolved the birds.

Modern birds are the dinosaurs' survivors. They are so close in structure to their small dinosaur ancestors that many biologists class them as dinosaurs, and what we traditionally regard as dinosaurs are now classed as "non-avian dinosaurs". For the purpose of our book, we shall continue to use the term dinosaur to denote these "non-avian dinosaurs" – the ones that became extinct at the end of the Cretaceous period.

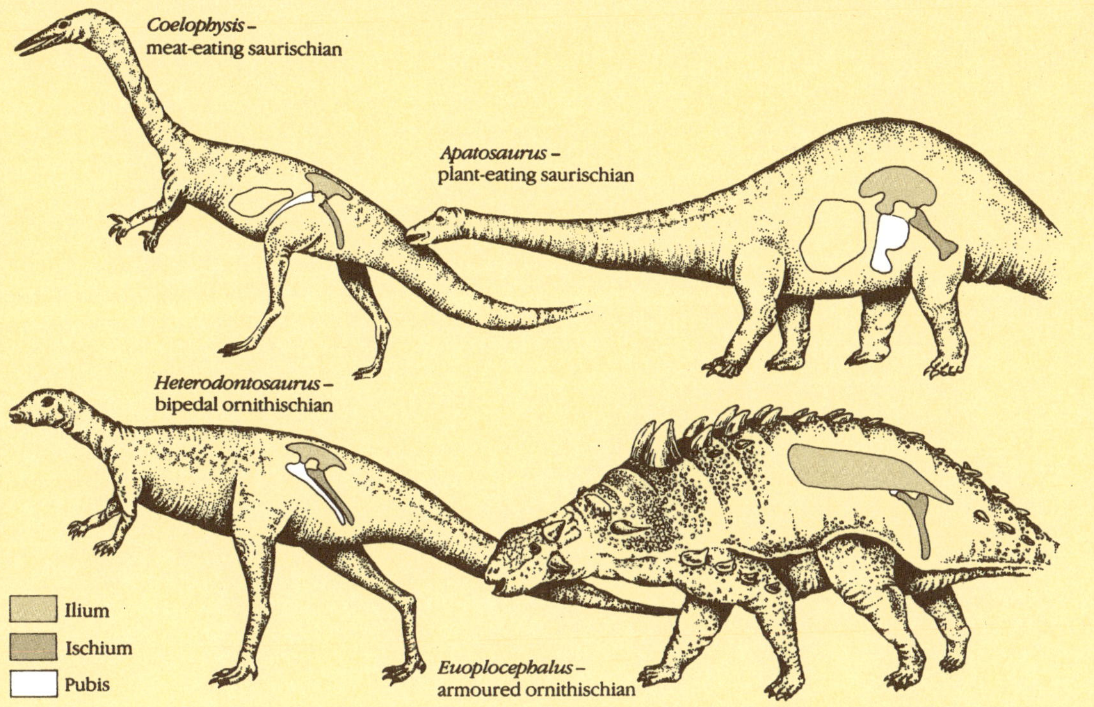

Dinosaur classification
Dinosaurs are classified on their hip types. The saurischians were the lizard-hipped dinosaurs, so-called because the arrangement of the bones of the hip, with the ilium attached to the backbone, the pubis pointing forward and the ischium pointing back, was similar to that of a lizard. Meat-eating saurischians walked on two legs. Heavy-bellied plant-eating saurischians walked on four. The ornithischians were the bird-hipped dinosaurs, with the pubis swept back against the ischium, like that of a bird. The ornithischians were all plant-eaters with the pot-belly slung beneath the hips. Many armoured forms had a greatly expanded ilium to support the weight of the armour.

The dinosaurs ruled the Earth for 170 million years, they dominated all other forms of life, and they were the most successful group of creatures that ever lived. What made them so overpowering, so dominant, so successful? Basically they were a very vigorous and constantly evolving group. As conditions changed throughout the Mesozoic era, they changed to accommodate them. Where there were deserts, there were desert-living dinosaurs. Where there were swamps, there were swamp-living dinosaurs. Forests produced forest-living dinosaurs. It is likely that there were mountain-living dinosaurs as well, although this is difficult to determine since mountain animals are rarely fossilized.

There is a strong possibility that many dinosaurs were endothermic; they had the metabolism that could allow them to regulate their body heat irrespective of the external environment. This is the condition that is commonly termed 'warm-blooded' and is found in the mammals and the birds. It produces a very high efficiency in an animal, and although a warm-blooded animal needs more food, this food can be more quickly converted into energy which can be used for longer periods. The swift, active, meat-eating saurischian dinosaurs were very probably warm-blooded, since only a warm-blooded animal could be alert and active enough to pursue the life style of these creatures. Most warm-blooded animals are covered in fur or feathers to ensure adequate heat regulation. The pterosaurs, the dinosaurs' winged relatives, were definitely warm-blooded and are known to have had fur. It is possible that warm-blooded dinosaurs may have had fur or down as well. It is difficult to say, however, whether the large plant-eating dinosaurs were warm-blooded. The argument against this is that the shape of the head and neck would have made it impossible for a long-necked plant-eater to consume enough food to sustain its body in a warm-blooded life style. The huge bulk of the body would probably have remained at a fairly constant temperature anyway without any complex warm-blooded system. However, sections cut through fossils of the bones of large plant-eating dinosaurs seem to suggest that they were fast-growing animals – and this characteristic is only found among warm-blooded creatures.

Structurally, the dinosaurs were very soundly built. Other reptiles have legs that appear to stick out at the side, with their bodies slung between. This is efficient enough for a small animal like a lizard, but not sufficient for supporting the bulk of a dinosaur. The dinosaurs evolved an upright stance, like that of a modern mammal. In other words, the legs were held vertically beneath the body, so that the weight of the body was carried at the top of the legs.

The skeletons of the largest dinosaurs were masterpieces of natural engineering. The legs were massive, to support the great weight, but the backbones were made of lightweight hollow struts, arranged so that maximum strength was provided by minimum bulk.

In our alternative zoology the Great Extinction did not happen. All these traits have therefore continued. The dinosaurs have evolved and adapted to the Earth's changing conditions. The dinosaurs, and the other great reptiles in the air and in the sea, are *still* the most successful and widespread animals on the Earth's surface. Now, before we explore the world of great reptiles as it is today, let us look at the history of animal life since the end of the Cretaceous period 65 million years ago. Let us explore the evolutionary development of the giant reptiles during the subsequent Cenozoic era.

The New Tree Of Life

The continuation of the dinosaur lineage over the last 66 million years has meant that these magnificent creatures have flourished, expanded and diversified with the same vigour as in the previous 150 million years. More so, in fact, since the last 65 million years have witnessed sweeping changes on the Earth's surface.

Most of the dinosaur lineages had actually died out long before the end of the Cretaceous period, and the dinosaurs that are alive today are descended from the most vigorous of the late Cretaceous groups.

The theropods – the meat-eating saurischians – belonged to a few distinct evolutionary lines. The most significant are the carnosaurs, the coelurosaurs and the ceratosaurs. They are distinguished from one another by details of the bone structure. The carnosaurs were the first to achieve a large size and included the allosaurs and the sail-backed spinosaurs. The coelurosaurs were the most varied, from little chicken-sized beasts to the mighty *Tyrannosaurus*. The ceratosaurs eventually gave rise to the huge abelisaurs that became prominent worldwide and are the biggest of the meat-eaters at the present day.

The heavy long-necked plant-eating saurischians, or 'sauropods', became abundant in the Jurassic period, but only existed in the late Cretaceous period in places where they were not ousted by the ornithopods. Those that survive today are mainly descendants of the titanosaurs and generally live on the southern continents.

The small bipedal ornithischians – the 'ornithopods' – (pages 14–15), like the small meat-eating saurischians, have adapted and flourished, adjusting to the great changes in environment with ease. Most of the plant-eaters today have evolved from this stock.

Ceratopsians – the horned ornithischians were very successful at the end of the Cretaceous and are still found in one form or another today.

The stegosaurs – plated ornithischians – had almost all gone by Cretaceous times and are extinct today.

Ankylosaurs heavily armoured ornithischians – were a successful Cretaceous group and have survived.

Then there are the flying pterosaurs. Despite competition from the birds, the pterosaurs have survived.

Of the spectacular swimming reptiles, the long and short-necked plesiosaurs, and the mosasaurs have survived, but the fish-like ichthyosaurs have become extinct.

And the mammals? They began in the Triassic period as small insectivorous creatures. By the end of the Cretaceous they were still small insectivorous creatures. They have had no opportunity to expand and diversify, and are small insectivorous creatures to this day.

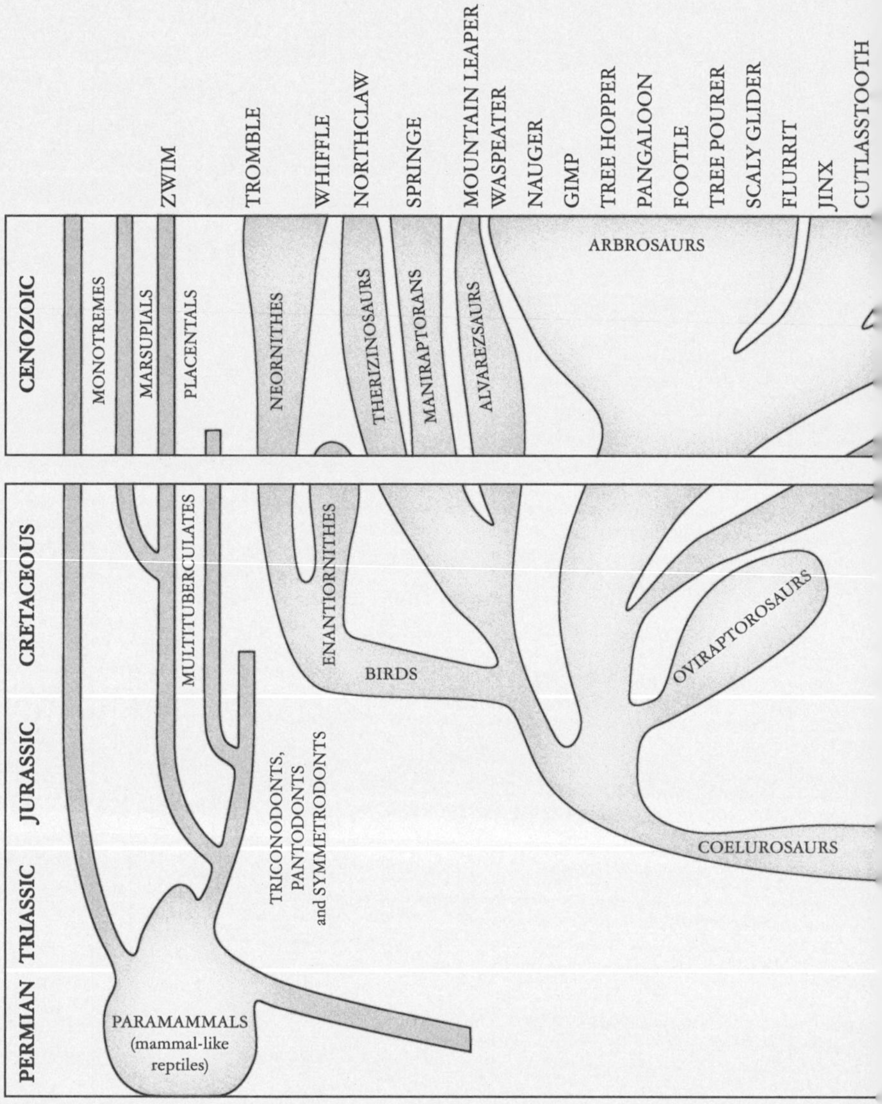

Maniraptorans
The maniraptorans were the small, lightweight coelurosaurs, a branch of which evolved into the birds. They have always been very versatile and, as well as the birds, they have produced all kinds of descendants. They are the ancestors of most of the small meat-eating dinosaurs of our new world.

Alvarezsaurs
These were a curious group evolved from the maniraptorans. Their arms were stubby and muscular and carried a single huge claw. In our new world they are restricted to a group of mountain-dwelling forms living in the Nearctic realm.

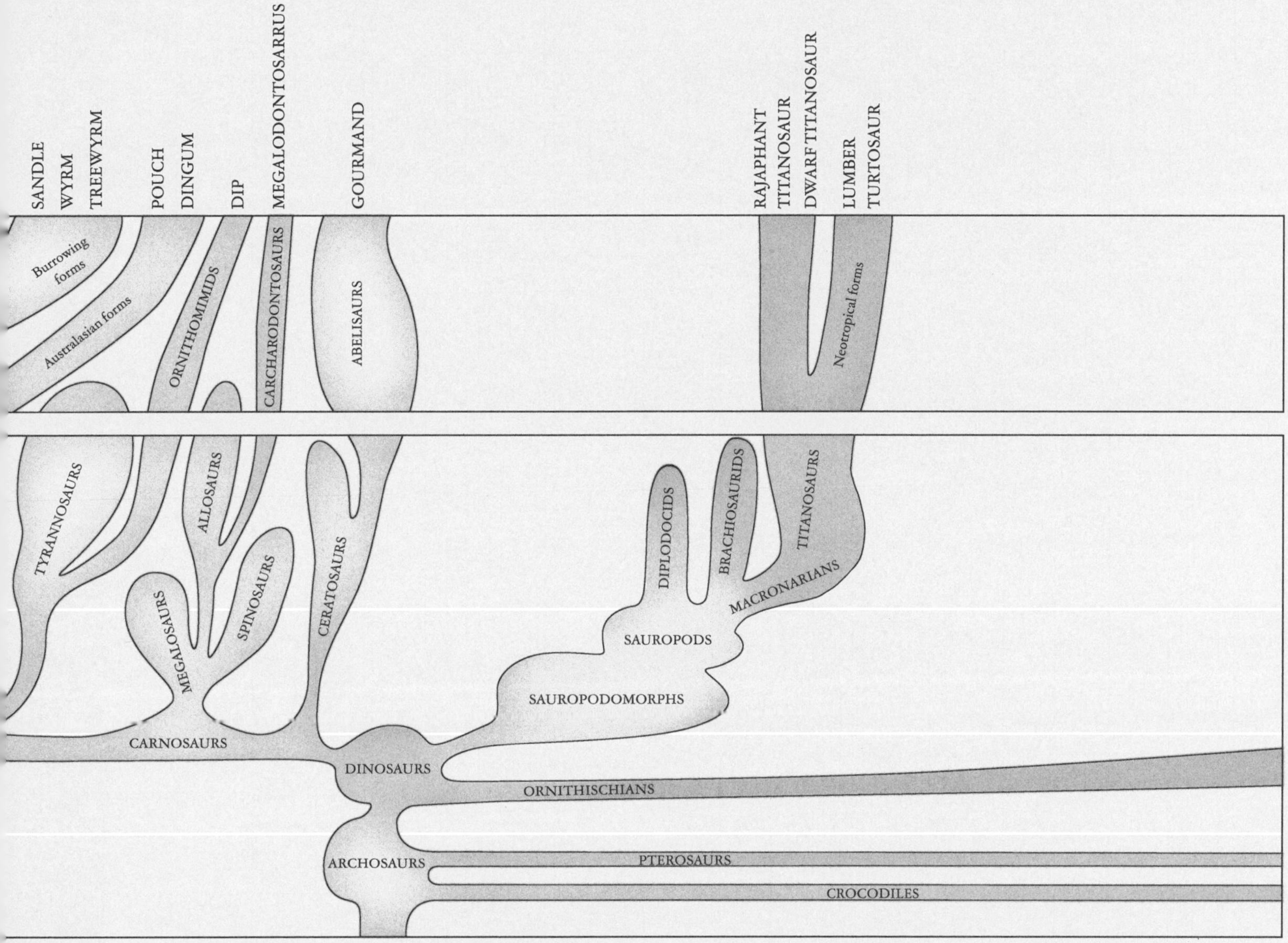

Therizinosaurs
This group of omnivorous theropods were quite widespread during Cretaceous times.
Now they there are only a few specialized carnivorous species found only in the Nearctic, where their huge clawed hands have developed into killing weapons.

Carnosaurs
The term carnosaur used to refer to any big theropod. Now it is restricted to the group containing the allosaurs and the huge carcharodontids of the Jurassic and Cretaceous periods. The latter still exist as the main meat-eaters, but only in a few isolated areas.

Abelisaurs
The abelisaurs evolved from the primitive ceratosaur group in Jurassic times and quickly became the biggest meat-eaters of the southern continents. Now they have spread to the north, where they have taken the place of the previous big meat-eaters, the tyrannosaurs.

Sauropods
The giant long-necked plant-eating saurischians – the sauropods – that dominated the world in Jurassic times are no longer so important. Their place as the main plant-eating dinosaur group has been taken by the ornithischians, particularly the ornithopods.

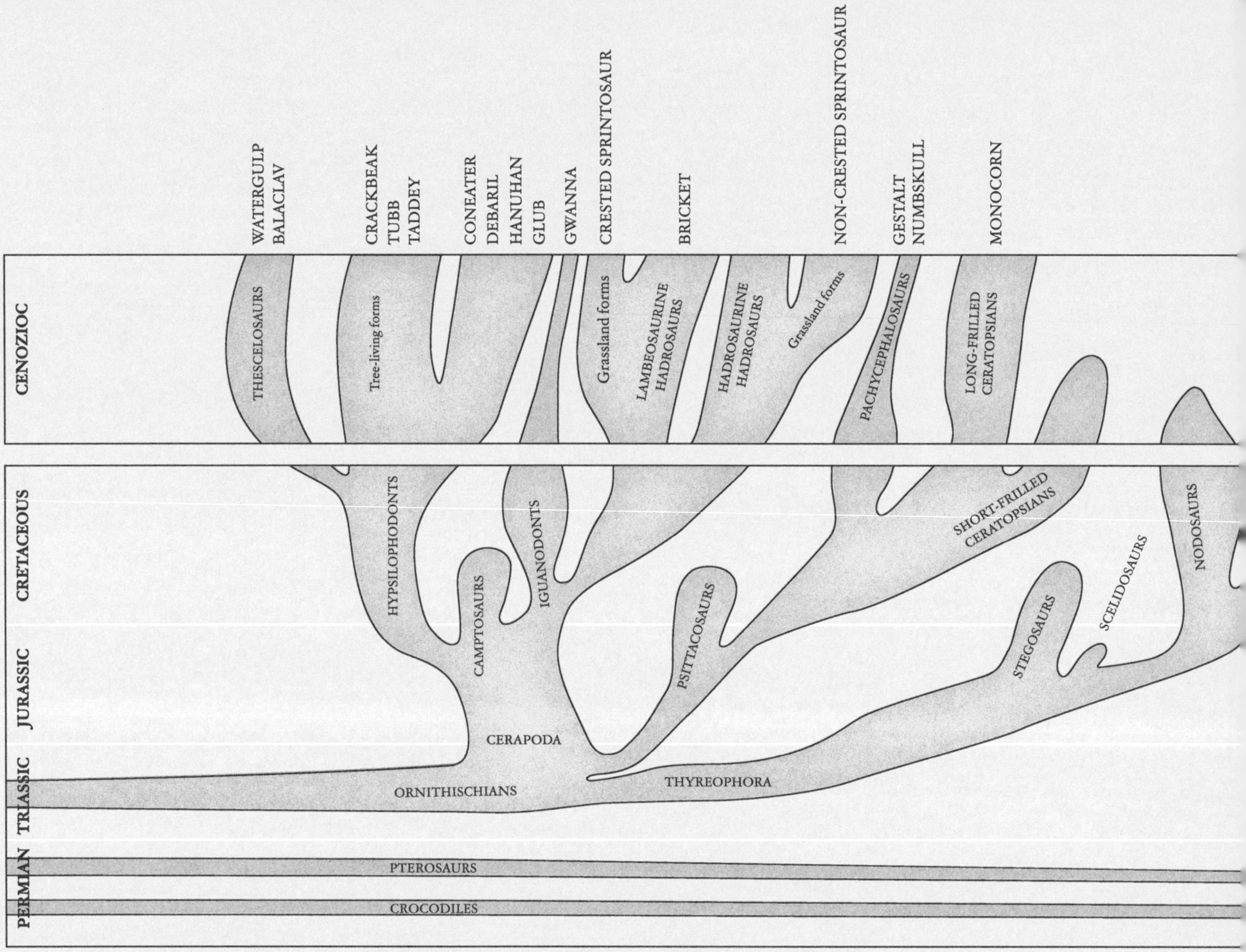

Hypsilophodonts
The hypsilophodonts were small swift two-footed plant-eating ornithischian dinosaurs. In size and appearance they were very similar to the early cerapods that appeared in Triassic times.

Iguanodonts
The iguanodonts were very much like the hypsilophodonts but tended to be much bigger and heavier. However there were quite small species as well, and it is these that survived to give us the present day examples.

Hadrosaurs
The hadrosaurs evolved from the iguanodont line and became the most abundant plant-eaters of late Cretaceous times. Their line split early into two major groups – those with flat heads and those with spectacular crests.

Pachycephalosaurs
Distant cousins of the hypsilophodonts and the hadrosaurs, were the dinosaurs with heavy armour on their heads. The first of these to evolve were the boneheads, characterized by the heavy bony growth on the roof of the skull, used for defence.

The COCONUT GRAB (page 99) and The KRAKEN (page 107) are invertebrates and therefore do not appear on this chart.

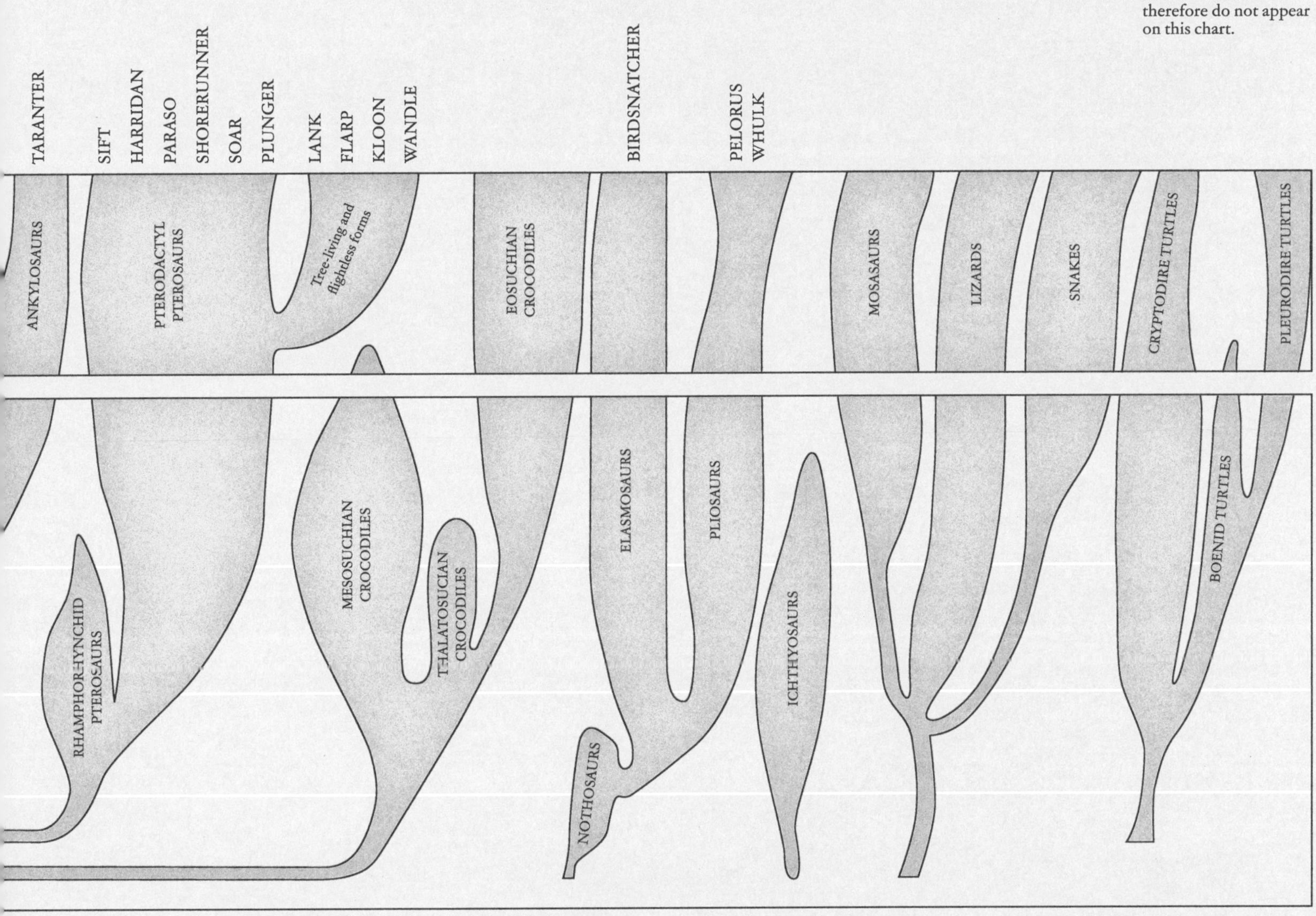

Ceratopsians
In the ceratopsians the armoured head was taken to an extreme. There was usually an armoured shield around the neck and an arrangement of horns on the face. Such an animal was very heavy at the front and so these were mostly four-footed.

Ankylosaurs
The ankylosaurs were the armoured dinosaurs that took over from their cousins the plated stegosaurs in Cretaceous times. There were two branches – the nodosaurids that had spikes on their sides and shoulders, and became extinct – and the club tailed ankylosaurids that survive today.

Pterosaurs
Distantly related to the dinosaurs were the pterosaurs, the flying reptiles. They evolved in the Triassic and are still abundant in the modern world. They compete successfully with the true flying dinosaurs – the birds.

Plesiosaurs
A number of different kinds of swimming reptiles exist in the oceans today. The most abundant are the plesiosaurs. Since they evolved in Triassic times there have been two main branches – the pliosaurs, with short necks and long heads, and the elasmosaurs, with long necks and tiny heads.

PALAEOGEOGRAPHY

THE EVER-CHANGING LANDSCAPE

Throughout the main part of the Age of Reptiles, the continents had been moving apart. During the Triassic period all the continents of the Earth were fused together into one great supercontinent called Pangaea, consisting of a northern section, Laurasia, and a southern section, Gondwana. During the Jurassic period this supercontinent began to split up, and in the Cretaceous period the separating continents were well on their way to their present-day positions.

Over the last 66 million years, since the end of the Cretaceous period, the greatest movement of continents took place in the southern hemisphere. At the very end of the Cretaceous, the continent of Australia was still attached to that of Antarctica. It subsequently split away and moved northwards until the continent reached its present position, in the southern tropical climatic belt. The continental block that now comprises India travelled northwards across the Indian Ocean, and it finally collided with the huge northern continent. As it split away from the original position as part of the east African landmass, it sheared away other continental fragments as well, and these have been left as the islands, known as Madagascar and the Seychelles, and a number of scattered submerged fragments. A seaway, called the Tethys, still separated the continents we now call Europe and Asia from that of Africa, but this seaway slowly closed and narrowed. At the other side of the globe the two continents of the Americas were quite separate, only connected by a string of islands during the earliest Tertiary times and with a permanent land bridge developing quite recently. North America, however, was almost permanently united with Asia across the Bering land bridge – a land bridge that has only recently become submerged.

The continental movement had an effect on the climate. With a continuous seaway around the world, made up of the Tethys and the gap between the Americas, there was a constant westward flowing equatorial current driven by the prevailing winds. This brought warm, moist climates to the edges of most of the continents. The water warmed in this current was swirled about to reach up along the coastlines in the north of the northern hemisphere and far down towards Antarctica in the south. Climates were warm and equable and humid forests grew on most of the continents.

As the continents moved, the Tethys Sea closed up. At the same time Australasia drifted away from the Antarctica and a circumpolar seaway opened up, allowing a continuous eastward-flowing current to sweep around the cold Antarctic continent. The equatorial current was therefore lost, and

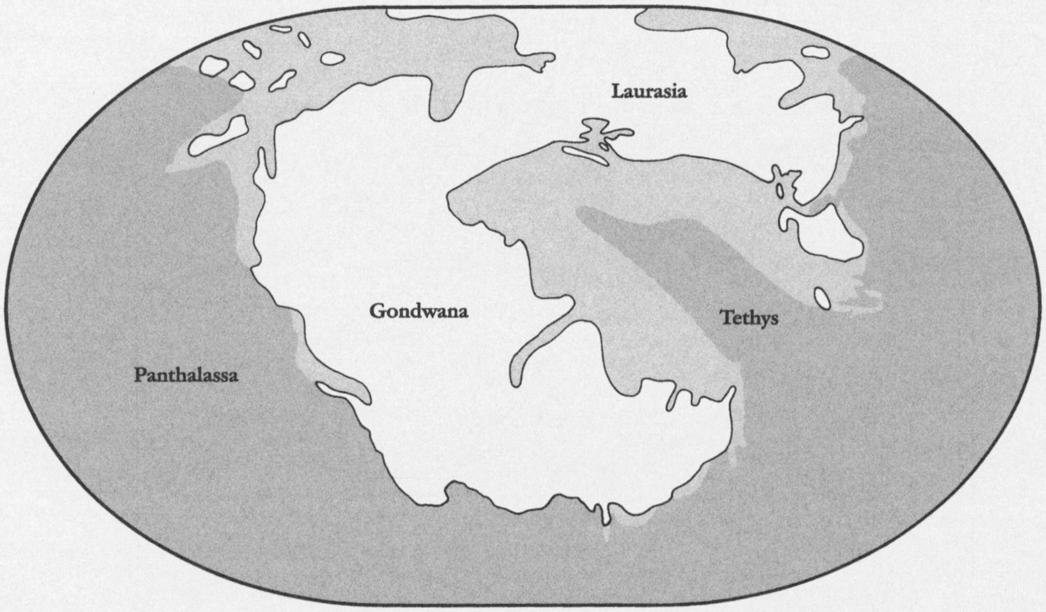

Pangaea
The Triassic period was the time when the supercontinent of Pangea was at its most complete. It was divided into two huge sections, each section far bigger than the greatest of today's continents. The northernmost part, consisting of modern North America, Europe and Asia, has been given the name Laurasia. The southern section, consisting of South America, Africa, India, Antarctica and Australia, is called Gondwana. An arm of the sea, called the Tethys, reached in from the east and almost separated the two sections. The rest of the sea area was united into a world ocean called Panthalassa.

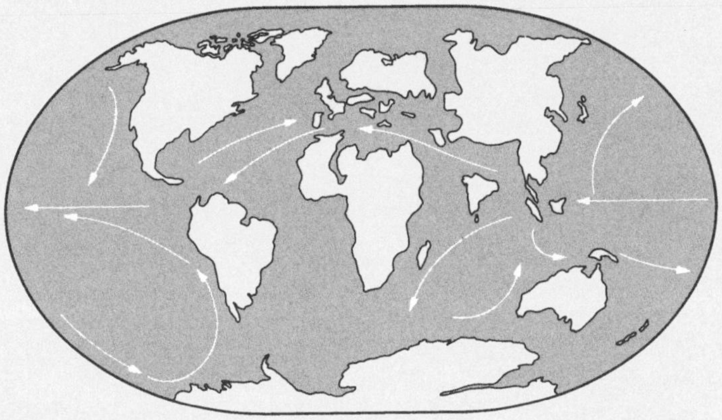

Early Cenozoic
As the elements of Pangaea separated, there was a clear seaway around the world near the equator, with a westward flowing equatorial current that modified the world's climate.

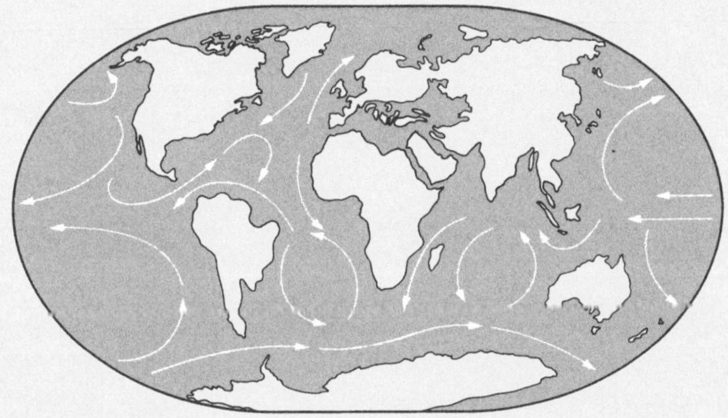

Late Cenozoic
Continued movements closed the equatorial seaway and opened up a circumpolar seaway around the Antarctic continent. Now the ocean circulation had no unifying effect on the global climate.

consequently the warm climates. There was now less mixing of warm tropical water and cool polar water, and climates increasingly differed over wide areas. The drier, cooler conditions that resulted meant that the lush tropical forests began to give way to grasslands – a new habitat.

The progressively cooler climates produced caps of ice at the poles of the globe. The Arctic ocean was nearly land-locked and the constant inflow from rivers diluted the salty water. As there was little mixing with warmer water from the south, this northern ocean froze over permanently. In the southern hemisphere the continent of Antarctica lay over the pole. This continent was far away from the warm ocean currents and effectively insulated from them by the circumpolar current. As a result Antarctica also froze over.

The climates continued to cool until the Ice Age, about 1.7 million years ago. Glaciers swept south from the polar ocean and down from the mountains, and the climatic zones were compressed upon one another towards the equator.

The topography of the world today is a direct consequence of this continental movement. When a continent moves, it tends to accrete mountain ranges along its leading edge. The greatest mountain chains of the Earth are the Himalayas along the join between India and Asia, where the two continents converged and collided 50 million years ago, and the Coast Mountains and the Andes along the west coasts of North and South America, where the westward movement is still taking place. The movement of the African continent against the European has created some contorted mountain chains in that area, producing the Atlas, the Appennines, the Alps and the other mountains that surround the Mediterranean. The string of East Indian islands between Asia and Australasia can be thought of as submarine mountain ranges, produced by the movement of submerged continental masses and ocean floors. Most of the other great mountain ranges of the world are old and worn away – relics of the continental movements of former times. The old mountains of the Appalachians along the east coast of North America were once continuous with the Scottish Highlands and the Norwegian Mountains along the edge of northern Europe, before these continents split apart. The Urals, the old mountain range between Europe and Asia, demonstrates where these two continents collided 300 million years ago. The movements that tore Pangaea apart are still at work. When a continent splits, it rises into a ridge, and cracks along the crest. The east African highlands represent such a rise, and the cracks are present as the Great Rift Valley that runs up the length of the continent to the opening fissure of the Red Sea.

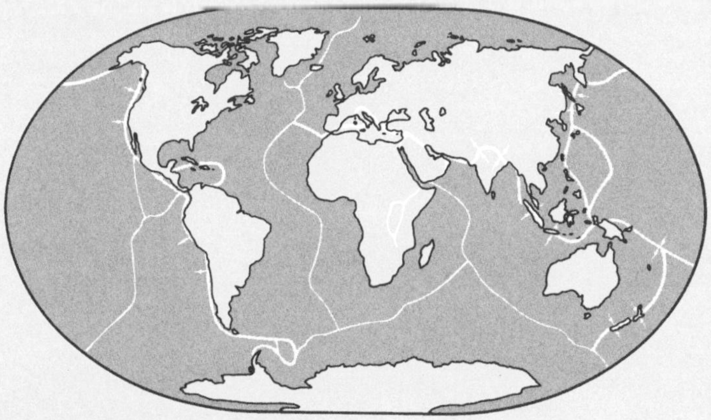

Modern plate movements
The Earth's crust is still in motion, and the plates are still continually shifting the continents. Along the active zones, where plates converge, mountains and volcanoes can be forced upwards.

Continental movements have a great effect on the wildlife. A continent may move from one climatic area to another, and consequently the animal and plant life has to evolve to adjust to the changing conditions. In another situation, the continent may move up against a neighbouring landmass and the two populations mix. All these effects are taken into account in the concept of zoogeography, that determines what assemblages of animals live in which areas of the world.

ZOOGEOGRAPHY

THE WORLD DISTRIBUTION OF ANIMALS

Animals are not the same the world over. There are a number of rules that define the kinds of animals that can live in one particular place and not in another.

The most obvious factor that determines the animal life of a particular area is the environment of that area – the climate, the terrain, the plants that grow there, other animals that exist there, in fact everything that contributes to the surroundings and habitat. We can see straight away that an animal that lives in a mountain environment, and is particularly well suited to live in a mountain environment, will not survive long if introduced into a swamp region – and vice versa. A mountain animal's tolerance of high altitudes would be useless in an area of low-lying wetland, and a swamp animal's stream-lined shape would be quite out of place on craggy mountain peaks. Yet, despite this profound influence of environment, a mountain animal may be quite closely related to a swamp region animal that lives on the same continent. They may both have evolved from the same ancestor just a few million years ago. On another continent on the other side of the world there may be mountain animals and swamp region animals that are likewise closely related to each other, but these will be quite unrelated to those on the first continent. The two mountain animals may look quite like one another, and the two swamp animals will share the same adaptations, but they will all have evolved quite independently.

Grouping the animals of the world according to their evolutionary relationships, rather than according to the environments in which they live, can divide the world into convenient regions known as ZOOGEOGRAPHIC REALMS. Each zoogeographic realm will contain a collection of animals that is peculiar to itself and that has evolved relatively independently from the animal assemblage of another realm. The boundaries between the realms may be well marked, as by oceans, or may be quite hazy, with several types of animals able to cross over from one to another. The existence of the realms is effected by the natural barriers to migration that arise because of the physical geography. A mountain range or a desert may divide one assemblage of animals from the next.

The ETHIOPIAN REALM consists of the majority of what we call the continent of Africa. The northern boundary effectively runs through the Sahara and Arabian deserts. Few animals can cross such a barrier, and the animals to the south have tended to develop in isolation from those to the north. The island of Madagascar is part of the Ethiopian Realm, although it is isolated from the continent and could be regarded as a little zoogeographic realm in itself.

The PALAEARCTIC REALM consists of the continents of Europe and Asia north of the Himalayas. It also includes the north coast of Africa. The Sahara desert represents a more powerful barrier to migration than does the Mediterranean Sea.

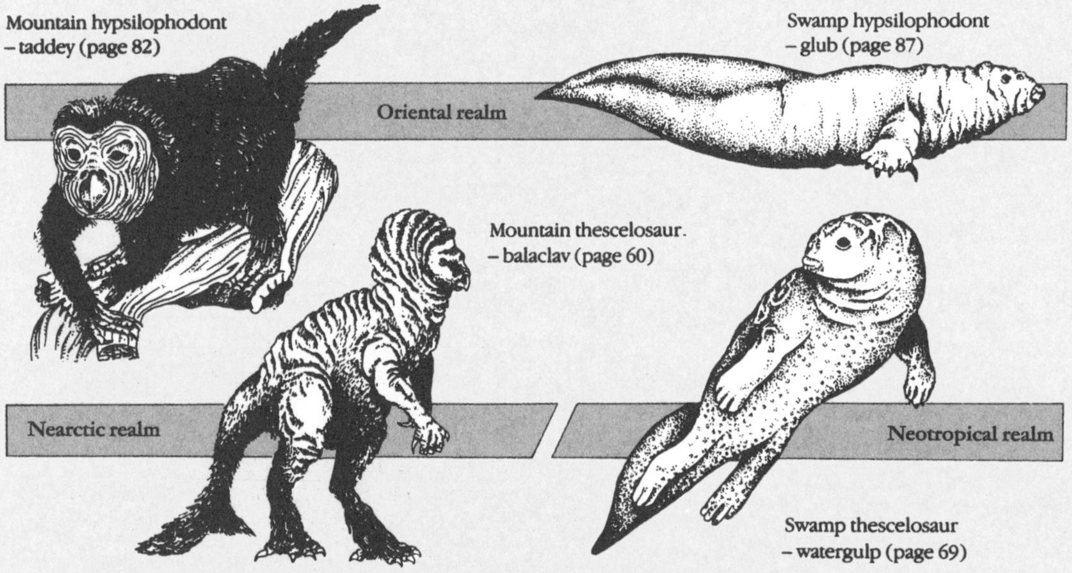

Environmental influence
During the Cretaceous the hypsilophodonts were very successful. A subgroup – the thescelosaurs – evolved in North America. The balaclav and watergulp are two thescelosaur descendants from the western realms. One lives in mountains, the other in swamps, and so physically, they differ greatly from one another. In the Oriental realm, the taddey and the glub are descendants of the main hypsilophodont group. The mountain form reveals the same adaptations as the balaclav, and the swamp form resembles the watergulp, but they are not closely related.

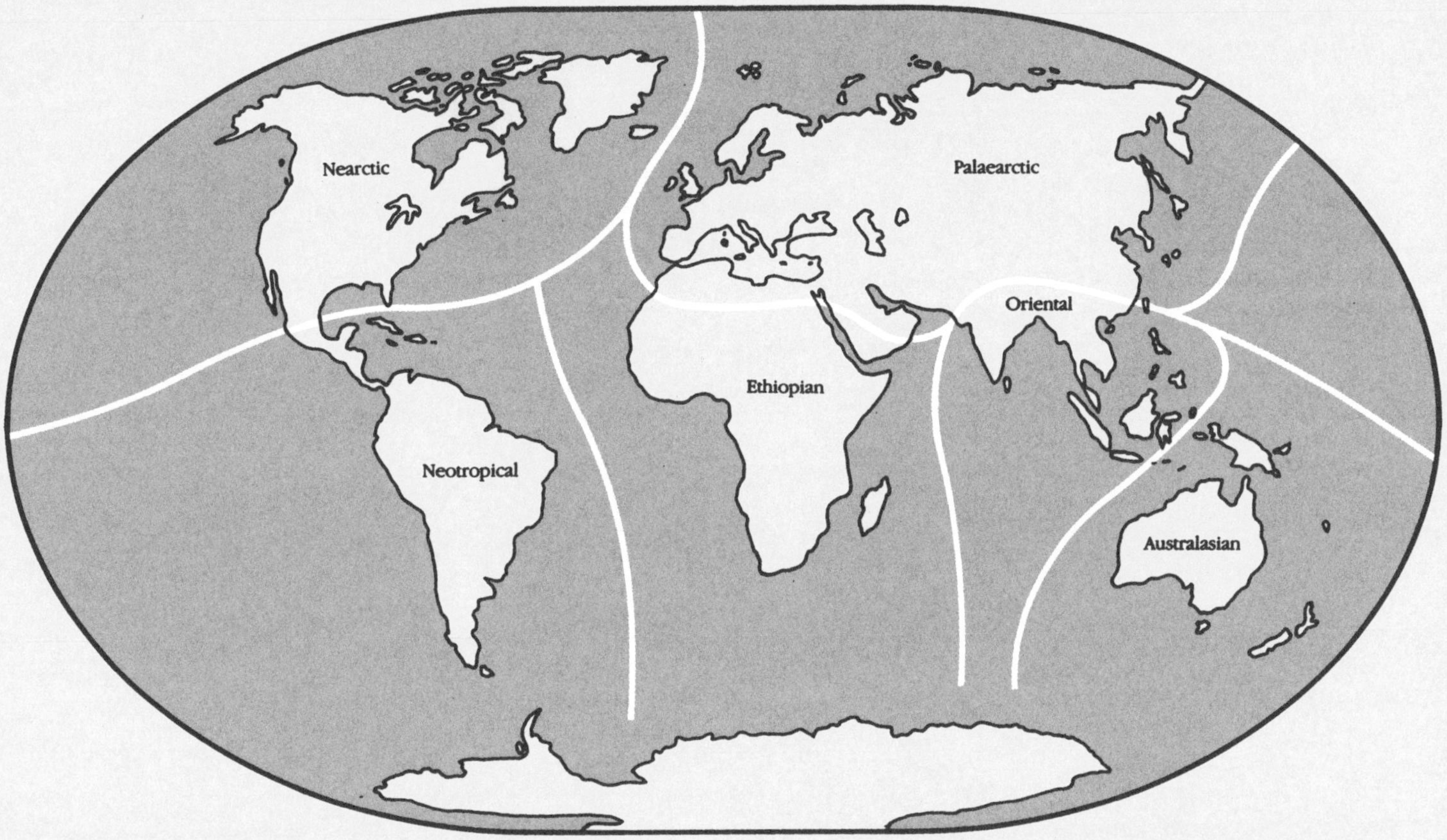

The zoogeographic realms
The landmasses of the modern world can be divided into six faunal zones, or zoogeographic realms, each realm with a completely different animal population. The boundaries between each realm are marked by barriers to migration and usually consist of deserts, mountains or seaways. The boundaries can be well-marked or indistinct. One great landmass on the globe – the continent of Antarctica – is so devoid of life that it is not contained in any of the zoogeographic realms.

The NEARCTIC REALM is the continent of North America, north of the Mexican desert. This desert isolates the realm from the continent to the south. The Bering strait isolates it from the Palaearctic realm to the north and west, but this is a very temporary barrier and the two realms share similar features.

The NEOTROPICAL REALM consists of the continent of South America and the island bridge of Central America. For most of the last 66 million years this has been a separate island, but quite recently it connected with the Nearctic realm and the creation of a land bridge has meant that the two realms are not as distinct as they once were.

The ORIENTAL REALM consists of what we call South-East Asia. The area of Asia south and east of the Himalayas, including most of the islands of the East Indies, comprises the Oriental realm. The mountains in the north and the desert in the west provide the barriers between this realm and the rest of the continent.

The AUSTRALASIAN REALM is perhaps the most truly isolated of all the realms. It has been an island continent ever since it was broken away from the continent of Antarctica and both plant and animal life there has developed in its own way.

To this list of six zoogeographic realms we have added a seventh, that of the OCEANS. Unlike the continents, the oceans are continuous and there are no physical barriers between one ocean and another. Animal life found in the oceans tends to be widely distributed and based upon temperatures and water conditions – in other words the environment – rather than the evolutionary development of particular areas. On the continents, the collections of animals found in each realm reflect the history of that realm. They depend upon when and how animal life migrated to that particular area and upon the subsequent development of environmental conditions such as geographic and climatic factors, that influence the animals living in that area.

In a world in which the dinosaurs and the other Mesozoic creatures survive, the animal life is still subject to the constraints of modern zoogeography. This book is divided accordingly. With no extinction of reptiles there has been no development of mammal life, and no evolution of human beings, and therefore no training of geographers. Hence the places of the world have not been given the names that they are known by today. Therefore, for the purposes of this book, we have dispensed with the common geographical names that are based on cultural and political divisions and reflect human exploration and history, and throughout we shall only refer to places in the context of their zoogeographic realms.

THE HABITATS

NATURAL ENVIRONMENTS OF THE WORLD

EQUATORIAL FOREST

The steamy jungles of the equatorial belt

Equatorial forests are found along the equator in the Neotropical, the Ethiopian, the Oriental and a corner of the Australasian realms. Along the equator, a region of atmospheric low pressure is created, where the air heated by the tropical sun constantly rises. Consequently, winds blow continually into this region from both north and south, bringing heavy rainfall. The ensuing hot, wet conditions are ideal for plant growth, and forests and jungle flourish.

Tropical rainforest
Different layers of the tropical rainforest support different animals. The emergents are home mainly to birds and pterosaurs. The canopy of interlocking branches contains most of the tree-living animals. Fewer creatures live in the gloom of the forest floor.

The hot, moist conditions are similar to those that existed during most of the Mesozoic era, and many of the Mesozoic types of dinosaurs live successfully here. There are, however, few large ground-living plant-eaters, since very little undergrowth survives in the gloom of the forest floor, and the tree trunks are often so close together that it is difficult for a big animal to move. The big animals are restricted to forest clearings and river banks. Most ground-dwelling plant-eaters are quite small and feed on roots, or seeds and nuts.

Insects abound on the forest floor, and many insectivorous animals live here. The ants and termites have evolved quite recently, and certain dinosaurs have evolved an ant-eating adaptation to exploit them.

There is no record of tree-living dinosaurs from the Mesozoic era; however, the tree-climbing habit has evolved among the new dinosaurs. These tree-dwellers are small, active and warm-blooded. The warm-blooded condition is necessary for an active tree-living existence that requires smooth muscular coordination. Tree-living dinosaurs have grasping hands and feet to enable them to grip branches, and they have evolved strong shoulder muscles, probably by re-evolving obsolete collarbones. The evolutionary lines of the coelurosaurs and the small ornithopods produced a number of tree-living dinosaurs.

Animals that jump between the branches of closely-spaced trees often evolve into flying and gliding creatures. Several gliding dinosaurs evolved from the tree-living coelurosaurs and these are found in the tropical forests in various parts of the world. With so many different varieties of tree present in the tropical forest, and so many different foods, such gliding animals do not compete directly with the birds and pterosaurs that also live there.

Collarbones
A conventional coelurosaur, *above*, has a very small collarbone, as the arms do not need to be very strong. Its descendant, the tree-living arbrosaur (page 33), *right*, has a strong collarbone to support its climbing muscles.

GRASSLANDS

The open plains

Grasslands are a relatively recent innovation to the face of the planet. There may have been areas of open country in the Mesozoic era but these would not have been clothed in grass. They are more likely to have been meadows of ferns and horsetails.

Tropical savanna
Grass is the important plant type on the open plains as it can survive long periods of drought, and can regenerate quickly after it has been grazed heavily. The only large plants are hardy thorn trees and scrubby bushes.

Tropical grasslands are now found in two parallel belts, north and south of the equatorial rainforest belt. As the sun's position in the sky moves north and south with the passing of the seasons, it brings heavy tropical rain in the summertime. During the winter the tropical rain is precipitated in the opposite hemisphere and dry conditions prevail. Trees are not suited to this seasonal weather but grasses thrive. Tropical grasslands are found in the Ethiopian, the Neotropical, the Australasian and part of the Oriental realms. Temperate grasslands cover large areas of the Nearctic and Palaearctic realms and are found in dry areas far from the ocean.

Grasslands, both tropical and temperate, represent a totally novel environment for the dinosaurs. Grass is a notoriously difficult substance to eat. It has a high content of hard silica, and produces a great deal of wear on teeth. It is also very indigestible, requiring sophisticated digestive systems. The geography of grasslands also imposes restrictions on the animals that live there. An animal roaming open plains can be seen from a long way off, so it is conspicuous to any hunter. A grass-eater must, therefore, be able to see any danger approaching from a long distance away, and be able to avoid it by running.

The duck-billed ornithopods of the late Cretaceous period have developed into grassland dwelling animals. Their batteries of teeth are ideal for coping with tough grass because they are constantly being replaced as they wear out. The development of a grassland animal's body, with its long running legs, does not represent a very great evolutionary step. Likewise, the long-necked sauropods that have survived on the southern continents, are mainly grassland-dwelling animals; they have not had to develop new adaptations because their sheer bulk is enough of a deterrent against the plains-living predators.

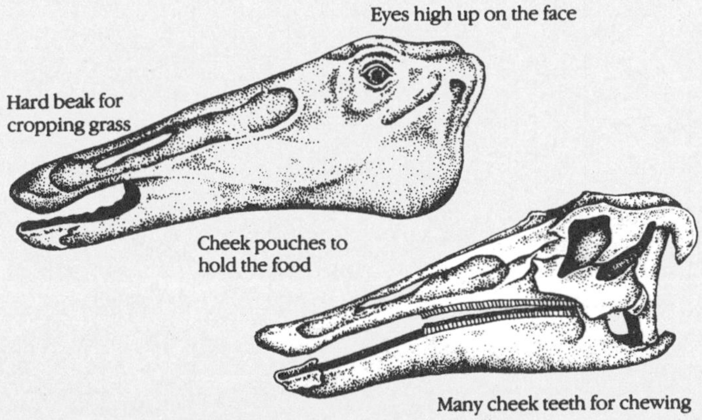

Grass-eating mechanism
The sprintosaur (page 56–7), a descendant of the duck-bills, has a long face with deep cheek pouches. When it grazes, its eyes are well above grass level, looking for danger. Its jaws have batteries of teeth that are constantly replaced as they are worn down.

DESERT

Hot dry wasteland

Desert conditions are found in regions where the climates are too harsh to support luxuriant vegetation, and hence do not provide a great deal of food to be exploited by animals.

There are several types of modern desert. The largest desert areas are the tropical deserts that lie along the tropics of Cancer and Capricorn. Here the hot air that rises at the equator – the air that drops all its moisture into the equatorial forests (page 20) – cools and descends. As a result, very little rain falls in these areas and few plants struggle to grow effectively in the constant dryness.

Continental deserts lie deep in the interiors of the largest landmasses, particularly in the Palaearctic and Nearctic continents. Here the sea is so distant that it has no effect on the local climate and the winds are always dry. Such deserts can be extremely hot in the summer, but icy cold during the winter months.

A rain shadow desert is found in the lee of a mountain range. The prevailing wind drops all its moisture as rain on the windward side of the range, and only dry winds pass beyond the mountains.

Since deserts are so inhospitable to animal life they often mark the boundaries between zoogeographic realms, and indeed provide the barriers that keep the realms distinct.

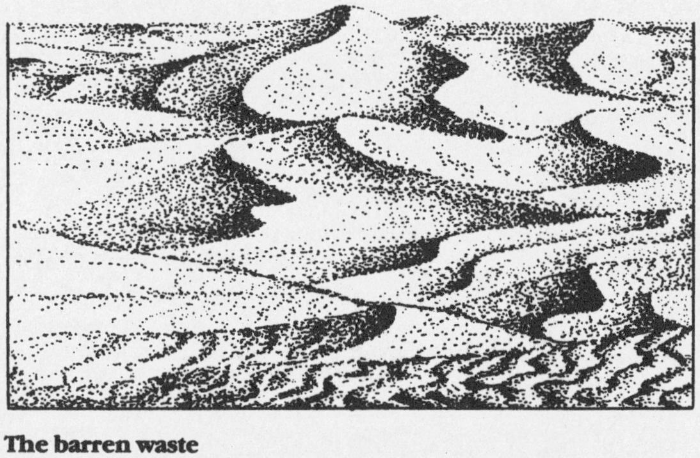

The barren waste
The desert surface can be quite varied. It may consist of flat wind-blasted rock, loose gravel, hard clay or salt, or constantly shifting dunes of sand as found in the tropical desert, *above*.

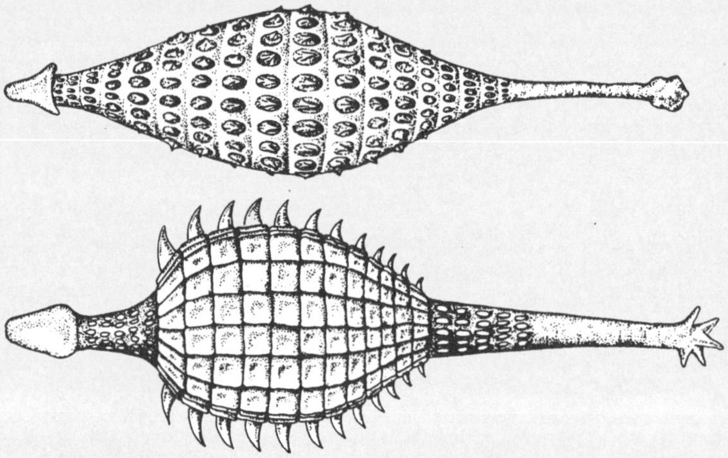

Water conservation
Desert animals need to conserve moisture. The taranter (page 52) uses the armour that it inherited from its ankylosaur ancestor, *top*. The armour has become a continuous waterproof covering *above*.

Desert conditions are so harsh that very few animals can survive in them. Compared with the equatorial forest, life is very sparse indeed, both in terms of numbers of species and abundance of individual creatures. Desert-living requires particularly extreme adaptations. Animals of the desert must be able to conserve their water to survive long periods in which there is little to eat. Plant-eaters usually manage this by storing all the available moisture present in the stems and seeds that they eat. Meat-eaters obtain water from the moist flesh of their prey. Both types of animals have very efficient kidneys that salvage every drop of moisture, and they rarely need to drink. Desert animals must also be able to protect themselves from the fierce heat of the day and the intense cold of the night. Small animals can do this by sheltering in rocky crannies or by burrowing in the sand. Many animals have evolved an impenetrable outer armour, not for defence, but to reduce the evaporation of their body moisture.

Many of the reptile groups evolved in desert conditions and so most of the desert vertebrates found today are reptiles. These include several dinosaurs, most of which have developed the special adaptations needed for desert survival.

TEMPERATE WOODLANDS

The habitat with four seasons

The temperate woodlands lie in the mid-latitudes of the Earth, part way between the tropics and the poles. These are regions of unstable climatic conditions, caught between the warm air masses pushing polewards from the tropics and the cold air masses pushing outwards from the poles. The boundary between the air masses is not a constant one, moving to and fro each season, and twisting and curling about from day to day. In the lower latitudes of this belt the western edges of the continents tend to have warm, dry summers and mild, damp winters, while the eastern continental edges are continuously warm and humid. Towards the poles, the cold polar air has a more noticeable effect and damp conditions are evident in both summer and winter.

Spring
In the spring the plants begin to grow. Most flowers and blossoms come out at this time and they are fertilized.

Summer
During the summer the main growth takes place, and the leaves soak up the sunlight, building up stores for later.

Autumn
In the autumn growth stops. The fruits ripen, allowing the seed to be spread. Most trees shed their leaves for winter.

Winter
Winter is the dormant season. The plants, and the animals, conserve their energy for the growing period ahead.

The greatest areas of temperate woodlands lie in the northern hemisphere. The corresponding latitudes in the southern hemisphere are largely occupied by ocean. Hence the Palaearctic and Nearctic realms contain most of the temperate woodland areas, while only pockets are found in the Neotropical, Ethiopian and Australasian realms.

Because of the seasonal nature of the climate, the temperate woodland tends to consist of deciduous trees; those that grow well and flower in the spring, mature in the summer, fruit and shed their seeds in the autumn, and drop their leaves and lie dormant in winter. There is usually a thick undergrowth of herbaceous plants in the temperate woodland, which dies back in wintertime. The annual fall of dead leaves keeps the soil enriched.

Seasonal changes of climate did not take place during the Mesozoic era and any dinosaur, or dinosaur descendant, living now has to contend with the difference in temperature between summer and winter. This presents no real problem for warm-blooded animals, and they can remain active through most climatic conditions. The warm-blooded dinosaurs thrive in this habitat. During periods of extreme conditions, the smaller dinosaurs hibernate, taking cover to sleep away the worst of the cold. The tree-living vegetarian dinosaurs are a good example; being small and active, they are able to hibernate in hollow tree trunks after gathering a store of food during the summer and autumn. Larger dinosaurs rely on their bulk to retain body heat and energy during cooler periods. Although they tend to become sluggish in the winter, they do not hibernate.

COLD FOREST

The northern girdle of conifers

The greatest uninterrupted expanse of forest in the world is found in the higher latitudes of the northern hemisphere. It forms a ring around the northern polar region. The heavy mass of cold air that lies over the north pole spreads outwards and southwards and governs the climate of the region. The climate tends to be very cold, with much of the water in the region locked up in ice for most of the year. The growing season here is very short, only 50 to 80 days, and the trees that grow must be able to withstand very poor conditions. The trees are mainly conifers, which are best suited to the cold climate. They have needle-shaped leaves that serve to reduce the amount of water lost through evaporation. Conifers do not drop their leaves in the winter, and so they are always available to provide the tree's nourishment when conditions allow it. The conical shape of the trees is ideal for shedding the great weight of snow that is likely to fall during wintertime. The trees grow slowly, and usually more than one year is needed before a fertilized flower will produce a cone with seeds. With no leaves falling, there is no leaf litter, and the cold conditions inhibit the decay of any dead matter that lies on the ground. The topsoil is consequently very thin and there is little or no undergrowth that is able to survive. Similar growing conditions are found high up in mountainous areas, and coniferous forests are found here as well.

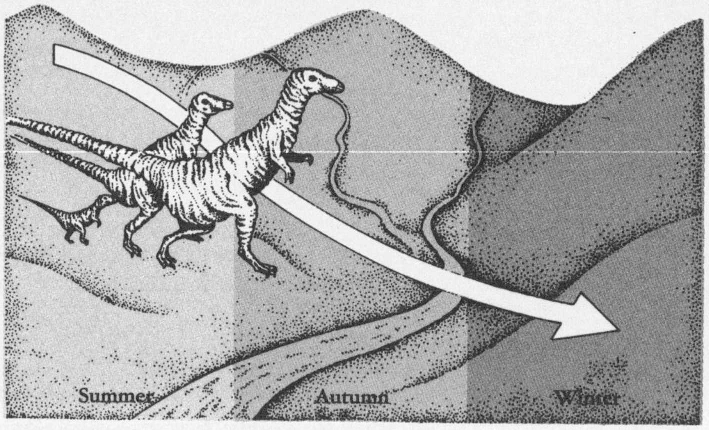

Migration
Migrating animals, such as the coneater (page 48), travel from the coniferous forest as soon as the snows begin, and spend the cold months by rivers in the temperate forest zone.

Coniferous forest
The thin soil of the coniferous forest floor is distinctly layered as the cold conditions mean there are few burrowing animals to mix it up. The trees often have a fungus coating their roots that helps them to absorb nutrients.

The great belt of northern coniferous forest lies in the Nearctic and Palaearctic realms, and mountain coniferous forests are found in small areas in all the other realms. There is no continuous coniferous forest belt in the southern hemisphere because ocean lies at the relevant latitudes.

Few animals exist in the cold forests as there is so little food available. Animals that do survive are very specialized feeders and do not compete with one another for food. Small animals, particularly mammals, are active during the winter, digging tunnels through the snow and feeding on seeds and insect larvae. The larger animals tend to migrate southwards to places where the climate is less harsh.

TUNDRA

The cold desert

Around the fringes of the northern ice cap, in the Palaearctic and Nearctic realms, there is a region of cold desert known as the tundra. During the long Arctic winter this region is as cold and bleak as the ice cap itself, but during the short summer the temperature may rise above freezing point, possibly attaining an average of 10 degrees Celsius (50°F). The covering of snow and ice will melt, but the water cannot drain away because of a permanent layer of frozen soil (the permafrost) below. The landscape becomes a waterlogged marshy waste of lakes and bogs.

A bleak landscape
The ground surface of the tundra thaws out in the summer, but the water cannot drain away because the ground is still frozen solid at depth. The summer tundra landscape is one of soggy marshes and temporary lakes with sparse vegetation.

The plant life of the tundra typically consists of low bushy herbs, mosses and lichens. At the southern fringes, straggly trees like birch and rowan grow as the tundra grades into the coniferous forest belt.

The growing season is very short, and many of the plants reproduce asexually rather than through the long process of flowering, fertilization and germination. The sudden growth of the tundra vegetation in the summer is accompanied by a bloom of insects that suddenly appear to take advantage of the brief period of warmth and sunlight. This short-lived flourishing of plant and insect life means that food is only available in the tundra for a limited period each year. Animal life in these regions tends to consist of migratory creatures. Birds flock to the tundra in the summer to feed on the abundant insects, but fly south for the rest of the year. Larger animals are also migrants, and wander the tundra wastes grazing from the mosses and lichens during the summer, but in the winter they are found in the south, sheltering among the great coniferous forests.

The areas of tundra that exist now are quite recent, from a geological point of view. They appeared when the ice caps appeared, reaching their greatest extent during the Ice Age. Evolution has not yet progressed sufficiently to produce a suite of animals that are specifically adapted to the strenuous conditions. There are hardly any modern dinosaurs here, despite the warm-blooded systems that they may have evolved. The environment is just too unpleasant. The large animals have evolved from the birds that have adopted a ground-dwelling existence because of the lack of predators; and these animals are only present during the summer. Throughout the year they migrate in herds north and south between wintering in the coniferous forests and spending the summer on the waterlogged tundra.

Annual migration
The large animals of the tundra, such as the tromble (page 50), spend their lives on the move. They migrate between summer pastures in the tundra, and winter in the coniferous forest.

WATER

The fluid surface of the earth

More than two-thirds of the planet is covered with water. The greatest depths are found in the oceanic trenches that are offshore along the mountainous edges of the continents. Most of the ocean floor consists of the abyssal plain, at depths of about 4,000 metres (over 13,000 ft). These regions contain living creatures but when compared with the shallower, more productive, regions of the ocean, they appear quite barren.

The continents are not entirely dry land, but are, to some extent, flooded at the edges. The submerged margins constitute what is known as the continental shelf, and this is usually shallower than about 150 metres (500 ft).

Sea water into which sunlight penetrates, and that is usually to depths of about 100 metres (300 ft), can support plant life. Sea-living plants are mostly algae, and can float near the surface or grow on the bottom in shallow waters. Small animals feed on these plants, and larger animals feed on the smaller. Dying creatures sinking into the depths feed other animals that live in the darkness.

Over a period of time, the greatest evolutionary changes tend to take place among the larger creatures. During the Devonian period certain fish left the water and so began the dynasty of land vertebrates. Not long after, several of the newly evolved land animals returned to the sea and re-adopted the aquatic way of life. The sea reptiles became large and spectacular during the Mesozoic period. None of these great reptiles were dinosaurs because the sea reptiles had become thoroughly established before the dinosaurs had evolved sufficiently to pursue that way of life. In modern times this pattern has persisted. There are no modern sea-living dinosaurs, but the traditional sea reptiles have continued to evolve and populate the sea.

To live in water it is necessary to have a streamlined shape. Water is a more difficult medium to move through than air. The classic teardrop shape found in fish is ideal, and an animal that chases and eats fish tends to evolve a similarly efficient shape. Paddles and fins are more use than legs and feet and these tend to be evolved, as do breathing systems that enable the animal to spend long periods beneath the surface. Many types of reptile have evolved these features to take advantage of the way of life offered by the sea.

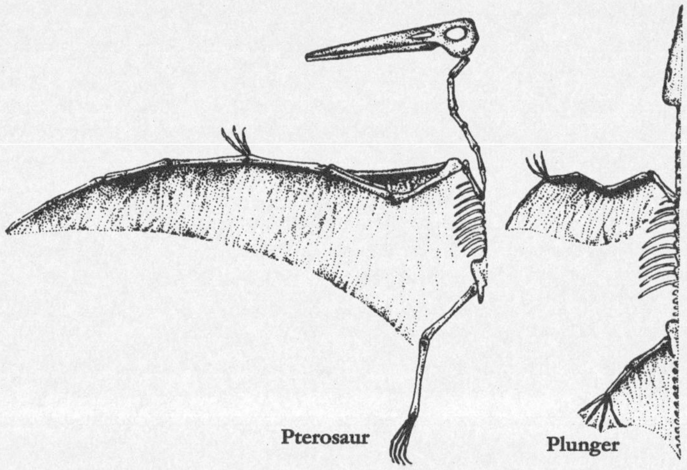

Swimming adaptation
The plunger (page 103) is a pterosaur that has changed from a flying to a swimming way of life. The body has become streamlined and fish-shaped. The wings have changed into steering paddles and a strong swimming paddle has developed on the hind legs and tail. It hunts fish in the euphotic zone of the ocean.

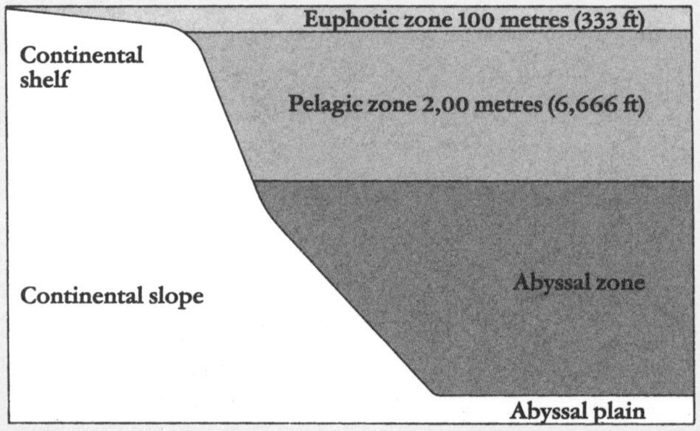

The ocean layers
Only the thin euphotic zone of the sea has abundant plant life. The pelagic zone is populated by hunters. Detritus feeders live deep in the abyssal zone, feeding on dead material from above.

AIR

The Earth's envelope of gas

Perhaps the most unusual habitat to be occupied by a living creature is the air. The colonization of habitats depends upon the availability of a food supply in that habitat. However, animals did not originally take to the air for this reason. Flight was a means of transportation from one area to another, whether from cliff-top to sea, or from tree branch to tree branch. Flight must have existed in the animal kingdom since the animals first left the water and came out on to land. Insects were among the earliest of the land animals and flight was developed early in their history. The earliest vertebrates to fly were various kinds of gliding lizard-like creatures in Permian and Triassic times. They flew by means of gliding membranes supported by extended ribs reaching out along each side of the body. There was no muscular action involved and the flight was no more than a glide that would not have been particularly well controlled.

The true masters of the skies became the pterosaurs. They were equipped with wings that were supported by their forelimbs and controlled by powerful muscles. They were fliers in the true sense and could flap their wings and control their actions very carefully. Many different types of pterosaur evolved, ranging from tiny creatures with a wingspan of a few centimetres that fed on insects, to huge creatures with wingspans of 12 metres (40 ft), that must have fed on carrion. The pterosaur's body became perfectly adapted for flight. The weight of a flying creature must be kept to a minimum, and the bones were hollow, dispensing with unwanted mass. The body must be rigid, to cope with the fantastic stresses put upon it by the action of flight. Hence, the back vertebrae were fused together as a solid rod. Quick reflexes and reactions are required to coordinate flying actions and keep the flight under control, and so pterosaurs developed a warm-blooded metabolism. As the Mesozoic period progressed, the pterosaurs were as successful in the skies as the dinosaurs were on the ground, and they have remained the masters of the skies today.

Halfway through the Mesozoic period, the birds evolved from coelurosaur stock. In many respects they parallelled the pterosaurs as they had wings, hollow bones, rigid bodies and a warm-blooded metabolism. They joined the pterosaurs in the air and now these two major groups share the environment of the skies between them.

More recently several dinosaurs have adapted to the air. They are most commonly found in the tropical forests where various gliding structures have evolved to help the tree-living dinosaurs move from one tree to another. These simple mechanisms are very primitive and quite similar to the gliding structures evolved by the early reptiles in Permian and Triassic times.

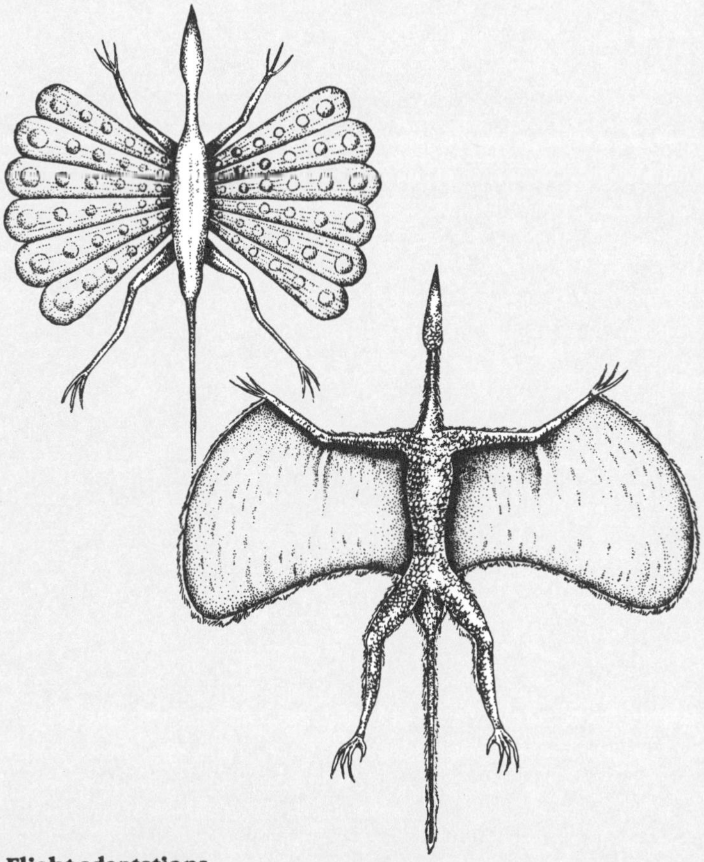

Flight adaptations
The gliding dinosaurs have adapted to flight in different ways. The scaly glider (page 71) of the Neotropical realm *top*, flies by means of paddle-shaped scales. The flurrit (page 85) of the Oriental realm *above*, has flaps of skin, or patagia, between its arms and body.

THE NEW DINOSAURS

The distribution of the continents and their environments in the world today is a legacy of the geological history of the planet, and of the current climatic zones. Each continent, or zoogeographic realm, has its own assemblage of wildlife. The particular group is determined by the flexibility of the boundaries to that realm, and by the past geological existence of those boundaries. During the Triassic and early Jurassic periods there was only one zoogeographic realm. The land areas of the world were united as part of the supercontinent of Pangaea and the same assemblage of animals was found everywhere. Since that time Pangaea has split up into individual continents, and remnants of the original fauna have developed independently in each continent. The animal life in a modern realm may be unique if the boundaries are precisely defined and impassable and have been so for tens of millions of years. On the other hand, the realm may reveal an assemblage that shares many species with its neighbouring realm, if the boundaries are ill-defined and easily crossed.

Upon the basic animal group in each zoogeographic realm are superimposed the environmental conditions. A single realm may contain tropical forests and deserts, chill tundra and ice-caps, and every environmental condition that would fall between. As these conditions are relatively new, having developed throughout the Tertiary period and especially since the Ice Age of the last two million years, the varieties of specialized animal on each realm have evolved from that realm's original animal assemblage.

The dinosaurs and pterosaurs that developed and established themselves during the Triassic, Jurassic and Cretaceous periods have now developed into browsing, tropical forest types; running and grazing grassland types; burrowing desert types; hibernating temperate woodland types, and migrating coniferous forest types. The shapes of the creatures are quite different from those of the Triassic, Jurassic and Cretaceous periods, but they have all evolved from the same original reptile stock.

In the pages that follow we explore the world as it is today. We look at the different zoogeographic realms and see how the combination of geological and climatic changes has allowed the evolution of the huge array of different types of animals. We view the vast and wonderful variety of modern animal life, and in particular, the new dinosaurs.

THE ETHIOPIAN REALM

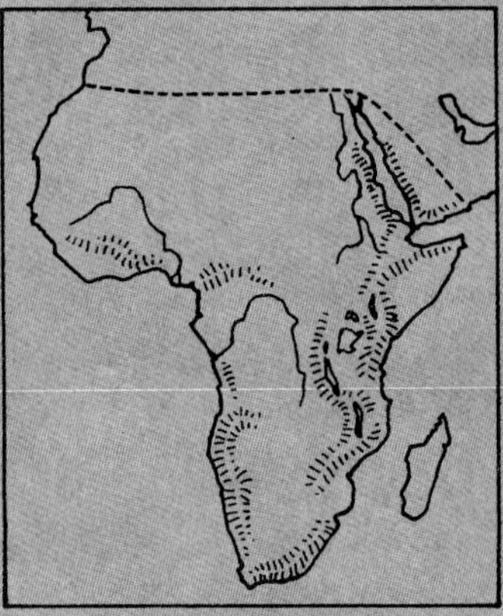

The continent that contains the Ethiopian realm is almost an island. Its broadest section lies north of the equator, where it stretches 7,000 kilometres (4,350 m) from east to west. South of the equator it narrows to about 3,000 kilometres (1,900 m) as it reaches to its rounded tip in the far south. From south to north the continent is about 7,000 kilometres (4,350 m) long. It is joined to the main northern continent only by a narrow isthmus in the north-east. The Ethiopian realm occupies the whole of this area except for a strip along the northern edge. The boundary that separates it from the Palaearctic realm in the north is the vast expanse of inhospitable desert that stretches across the continent's widest portion.

The Ethiopian continent was once the central portion of the great southern supercontinent of Gondwana (see page 16). In Triassic times its north-western edge provided the join between Gondwana and the northern supercontinent called Laurasia, uniting the two in the all-encompassing super-continent called Pangaea. These two supercontinents became separated during early Jurassic times, by an arm of the sea called the Tethys. Then Gondwana began to break up. The block that was to become the Ethiopian continent did not shake itself completely free from its neighbours until the late Cretaceous period, and since that time it has been moving northwards across the Tethys. The Tethys is gone now - closed up by the northward movement of the southern continents. A large island, about 1,200 kilometres (745 m) long, lies about 700 kilometres (430 m) away off the south-eastern coast. This is a mini-continent in itself - another fragment of Gondwana - and has been isolated for so long that it could constitute a tiny zoogeographic realm on its own.

The most mountainous regions lie along the eastern edge of the continent. Here there are mountains that are so high that their summits are snow-covered all year round despite the fact that they lie almost on the equator. Along this edge the forces that tore Gondwana apart are still at work. A rift valley, formed by tension in the Earth's crust, stretches from south to north and shows that the eastern part of the continent is pulling away. Earthquakes and volcanoes are common here, and at some time in the future the whole eastern section of the continent may drift out to join the other continental fragments that are scattered across the ocean to the east. At its northern end this rift valley is already submerged and has formed an inlet of the sea. Other mountains are found along the western coasts. These are the eastern edges of old rift valleys that appeared during the days of Gondwana when the continent began to rip itself away from the surrounding landmasses. The rest of the continent consists of flat river basins and plateaus.

The habitats of this realm are quite varied, and are arranged in a kind of concentric pattern. The central portion, covering the lowland areas through which the equator passes, swelters as a tropical rainforest. The prevailing winds bring constant rain to this hot area, and the water produces vast rivers that drain from the jungle basins into the sea. This is a zone of small tree-living animals, feasting on the myriad fruits and insects of the lush greenery.

To the north, south and east the forest grades into grasslands, where it is still hot, but the rain only falls for one season of the year. Trees do not do well under such conditions but grasses do, and open grasslands and sparsely wooded savanna sweep across the broadest part of the continent in the north, cover the foothills of the eastern mountains, and curve down to the southern hinterland. Long-legged grass-eaters live here, including some strange developments of flightless pterosaurs.

To the north and the south lie the deserts - harsh, hot and arid habitats. These are regions that lie along latitudes that receive only dry winds and little rain. The desert in the north, crossing the greatest width of the continent, is so inhospitable that it provides the barrier which effectively isolates the Ethiopian realm. The desert in the south is less extensive but no less antagonistic to life. Yet in the deserts there are animals - animals that have adapted to survive the waterless conditions. Here live tiny, legless, burrowing dinosaurs that have evolved to withstand the extreme killing conditions, and have gone on to be even more successful in less harsh environments.

The offshore island was once part of the same realm but must now be regarded as an individual environment. With tropical forest and open woodland it still supports the kind of animals that existed on ancient Gondwana before it shattered and scattered. Conditions there have not altered significantly since it became isolated and the animals have therefore not needed to adapt to a changing environment.

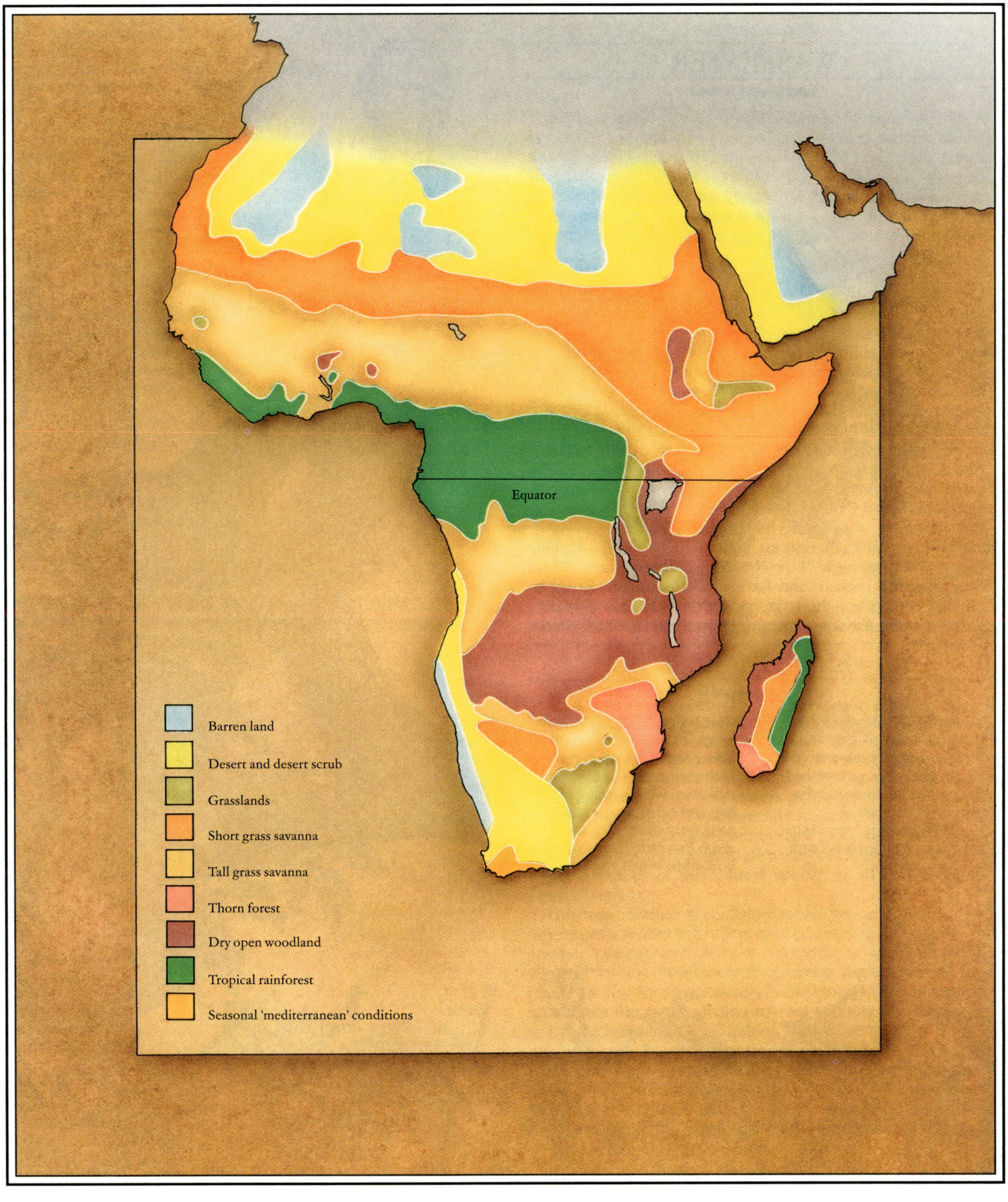
Equator
Barren land
Desert and desert scrub
Grasslands
Short grass savanna
Tall grass savanna
Thorn forest
Dry open woodland
Tropical rainforest
Seasonal 'mediterranean' conditions

Tropical rainforest

WASPEATER

Vespaphaga parma

A whole range of tree-living dinosaurs evolved and established themselves in the branches and boughs of the world forest after the end of the Cretaceous period; they rapidly diversified into all kinds of different forms. About that time the social insects – the bees, wasps and ants – began to develop. These live in colonies, usually consisting of an egg-laying queen sustained and protected by an army of workers and soldiers.

The waspeaters were a specialized branch of the tree-dwellers known as the arbrosaurs, that developed at the same time and preyed upon the social insects. After the world forest was broken up by the changing climates and the spread of the grasslands halfway through the Tertiary period, the waspeaters became restricted to a number of tropical forest areas. Most species now live in the equatorial regions of the Ethiopian realm. Their long claws are an adaptation to a climbing way of life, and are also useful for ripping at wasps' nests. Their scaly hide has grown into a roof of overlapping plates, impenetrable to the stings of the insects. Similar wasp-eating animals have evolved in the tropical forests of the Neotropical and Oriental realms. Many of these are related to the Ethiopian waspeater, having become isolated in the various tropical forest regions. Others, such as the pangaloon *Filarmura tuburostra* of the Neotropical forest (page 68) are only distantly related, and have evolved similar shapes by parallel evolution – the independent development in related animals of similar adaptations to allow them to follow similar lifestyles.

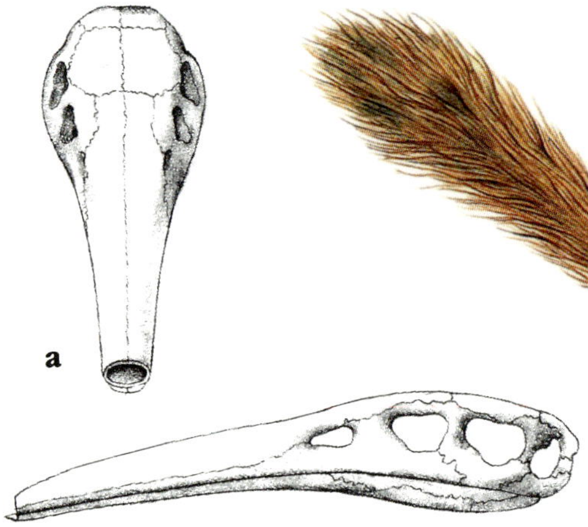

▽ An arbrosaur, such as the **tree hopper**, is agile and perfectly at home in the trees, but is a very ungainly animal on the ground. When it is forced to, it can only move over the ground by a series of undignified leaps, keeping its long arms and heavy tail well up out of the way.

▷ The bones of the **waspeater's** skull (**a**) have become fused into a narrow armoured tube that can penetrate deep into the nests of tree wasps. The strong hind legs and tail allow the animal to hang from the branch and reach into nests that have been built in awkward positions.

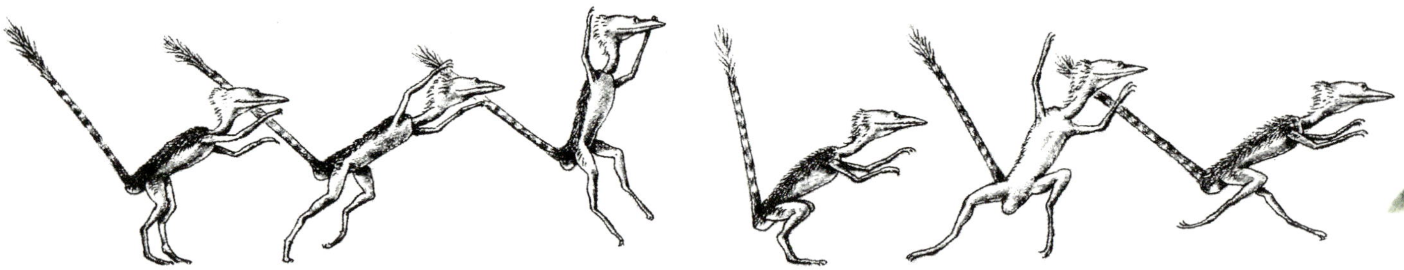

▽ Most meat-eating dinosaurs have lost their collarbone. They do not need one since the forelimbs do not need to be particularly strong. One or two forms do possess this bone, however, and it is not clear whether their ancestors retained it from the time when they went about on four legs, or whether it 're-evolved' having been first lost – a process known as secondary development. In *Arbrosaurus* and its relatives, the collarbone provides the anchorage for the very powerful arm muscles with which the animal can brachiate, or swing with an overarm action, through the branches of the trees.

△ An arbrosaur's tail is a stiff, straight rod. It uses it for balance when leaping about among the trees. The long claws on the three main toes and the three fingers are useful both for finding purchase on branches and for ripping up bark for insects.

Tropical rainforest

TREE HOPPER

Arbrosaurus bernardi

One of the most widespread developments in the dinosaurs after the end of the Cretaceous period was the evolution of the arbrosaurs. These evolved from the small coelurosaur theropods – the lightly built running flesh-eating dinosaurs. In Jurassic times the coelurosaurs gave rise to the birds, and the development of the arbrosaurs from the same stock was brought about by similar evolutionary processes. Perhaps the most typical of the modern arbrosaurs is *Arbrosaurus* itself. Various species of this animal appear in all the zoogeographical realms.

The main feature that distinguishes the arbrosaurs from other coelurosaurs is the presence of a strong collarbone. In another coelurosaur offshoot, retention of the collarbone allowed for the development of strong flight muscles and paved the way for the evolution of the birds. In the case of the arbrosaurs the collar girdle provides support for the strong arms which are used for climbing and swinging about among the branches. Its endothermic (warm-blooded) physiology enables it to pursue an active hunting life style. Its skull shows adaptations for this, with the big brain box, the eye-sockets directed forwards giving stereoscopic vision, and the narrow, finely toothed jaws – ideal for winkling insects out of crannies in tree-bark.

△ The **lank** runs with a pacing motion – both legs at the same side moving in the same direction at the same time. This prevents the long legs from becoming entangled.

◁ The long face of the **lank** means that its eves are still above the level of the grass even when its snout is grazing at the grass roots. It can thus keep an eye open for danger. On the open plains danger can be seen coming from afar, and the lank's long legs give it the speed needed to escape from predators.

△ It is difficult to believe that the **lank's** front legs are evolved from pterosaurs' wings. The fourth finger, that once supported the flight membrane, now carries the animal's weight and has a hoof on the end. The three small claws that once acted as a hand are now only used for grooming the animal's fine fur.

Tall grass savanna

LANK

Herbafagus longicollum

The tropical grasslands, when they developed and spread halfway through the Tertiary, were a totally novel environment for the great reptiles. Grass is a remarkably tough substance, and a grass-eating animal needs a number of extreme specializations to allow it to survive. Grass is full of hard silica and so it causes a great deal of wear to teeth. A grass-eating animal needs teeth that are constantly growing or frequently replaced. Complex digestive systems are needed to break down and extract nutrients from the substance of grass. On top of all that, a big grassland-living animal needs to have long legs to allow it to run away from danger. In some places, such as the prairies of the Nearctic realm, dinosaurs developed quite happily into grass-eating forms. However, on the savannas of the Ethiopian realm, dinosaurs were not able to establish a foothold. They were beaten to it by their cousins - the pterosaurs. A group of pterosaurs abandoned their powers of flight as soon as the tropical grasslands spread, and they became the main grass-eating animals of the Ethiopian realm, The lank is now the most highly specialized of these, and the one that has become most unlike its flying ancestor. The body has remained short but the forelimbs and hind limbs have become long and equal in length. The neck and face are also long - all plains-living features.

Tall grass savanna

FLARP

Vexillala Robusta

Another of the ground-dwelling pterosaurs of the Ethiopian savanna is the flarp. Its specializations are not so extreme as those of the lank, and its flying ancestry is still evident from the presence of vestigial wings on its forelimbs. It can be thought of as an intermediate evolutionary stage between a conventional flying pterosaur and something as specialized as the lank. Although it lives in the same environment as the lank the two animals do not compete for the same food. Being much smaller, only about a metre (3 ft) high at the shoulder compared to the lank's head height of 3-4 metres (10-13 ft), it feeds closer to the roots of the plants. The lank tends to graze the tops of the grass heads and leaves. The flarp's sharp teeth at the front of the mouth enable it to root about among the plants at ground level. It has fleshy lips and copious cheek pouches enabling it to chew the grass and plants thoroughly before swallowing. After digesting for a while in the stomach the grass is brought up again for more chewing before being swallowed for a final time. In this way nutrients are efficiently extracted from the plants. The flarp runs about the plains in small flocks of about a dozen and can often be seen displaying their brightly patterned wings, or squatting down in the grass in groups during the heat of the day and chewing over the morning's meal.

The structure of the **flarp's** forelimb and hand is similar to that of a flying pterosaur. The fourth finger is long and supports a flap of skin — the vestige of the flying wing. Now the wing flap is used for display - for attracting mates or for warning rivals. The fourth finger and the flap are normally held back out of the way **(a)**, but when displaying they are extended and the bright patterns are shown **(b)**. This display is accompanied by raucous screeches that carry for great distances across the plains. Flarps feed on plants and grasses that grow close to the ground. They have short, sharp teeth positioned at the front of the mouth **(c)**.

Desert and desert scrub

SANDLE

Fususaurus foderus

Hot deserts are found along the latitudes of the tropics, where air that has risen and lost its moisture at the equator descends and settles dryly over the land. In the Ethiopian realm the deserts are found in the south-west corner of the continent, and all across the north. The northern desert is so vast and inhospitable that it provides the boundary between the Ethiopian and the Palaearctic realms. Nevertheless, despite the harsh conditions, there are some animals that do live here. The coelurosaurs have proved to be adaptable enough to develop forms that are uniquely suited to such an environment – forms such as the sandle and the wyrm.

The days are so hot and the nights so cold that any animal must spend much of its time protected from the extremes. The sandle does this by burrowing in the sand. Its streamlined spindle shape is ideal for a burrowing way of life, since sand grains slide past its smooth contours easily as it tunnels through the dunes, digging and pushing with its stumpy legs. The eyes and nostrils are high up on the head, enabling it to see about and breathe while the rest of the body is buried. Desert animals need to be very careful about water conservation. The sandle never drinks but obtains water from the moist flesh of the animals it eats. Its kidneys are very efficient, allowing almost all the moisture to be retained, and secreting any poisonous compounds in a saliva that helps to paralyze and subdue its prey.

Desert and desert scrub

WYRM

Vermisaurus perdebracchius

Like the sandle, the wyrm is a streamlined animal adapted to a burrowing existence. It has also evolved from the lightly built coelurosaurs of the Cretaceous period. Its streamlining takes the form of a great elongation of the body and neck and the total loss of the forelimbs. Its head is trowel-shaped and used to force a path through the sand. There it moves by undulations of the long body and by thrusting with a swimming motion of the broad hind feet.

There are many species of wyrm, not all of which have continued to be burrowers. Many live in the burrows of other animals, such as the small desert mammals. Many others have spread into different habitats and different zoogeographical realms around the world. The long sinuous form that first evolved as a burrowing mechanism has applications in other ways of life. There are swimming wyrms in various parts of the globe and a whole range of tree-living wyrms, especially in the Oriental realm (page 84). The wyrms of the desert of the northern Ethiopian realm, where they first evolved, are still burrowing animals, and hunt small mammals and reptiles usually at dusk and at dawn. Like the sandle they obtain all their moisture from the flesh of their prey, who in turn obtain all theirs from the scrubby desert plants and seeds that they eat. They make very efficient use of their water despite the fact that they have only one kidney, the other having been lost as the narrow shape evolved.

△ The **sandle** has come a long way, in evolutionary terms, from the agile bipedal, long-tailed coelurosaur that was its ancestor. The body has become streamlined and covered in smooth fur. The clutching hands on the forelimbs and the strong running hind legs have become digging shovels. The scaly head is flat and chisel-shaped, ideal for thrusting its way through the sand as the tough scales prevent abrasion.

▽ The back of the **wyrm** is covered with fine hair. The rest of its body is armoured by a network of scales. The head is a scaly shield (**a**). A row of heavy transverse scales runs the full length of its underside (**b**) protecting its belly from the constant abrasion of sand and stones as it winds its way along. A heavy shield of scales covers the rump (**c**), to protect it from members of its own species that may pursue it down the tunnels of its prey.

c

a

b

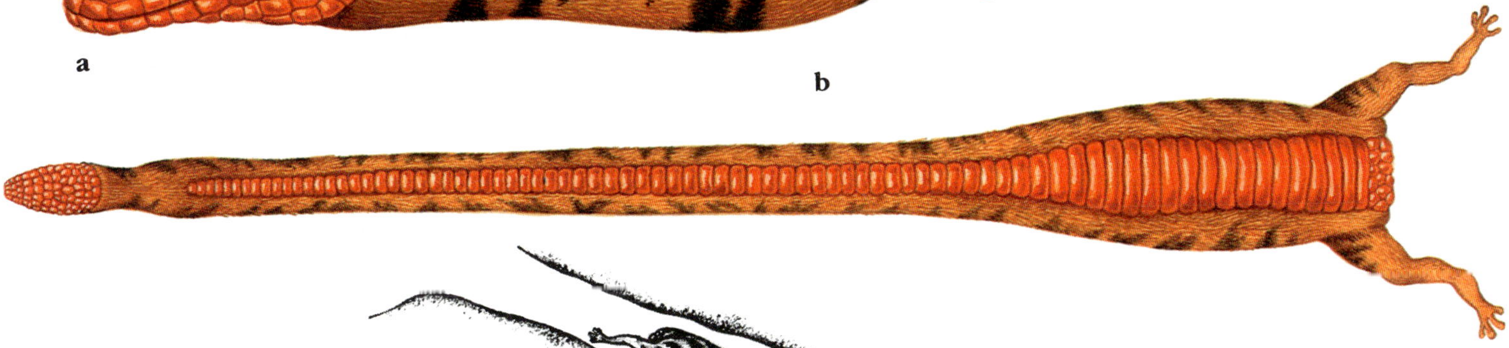

◁ The **wyrm** is long enough and narrow enough and agile enough to chase its prey down their burrows. Its diet consists mostly of the small hopping mammals that have evolved to live in the desert.

▽ The **sandle** feeds on small vertebrates, insects and scorpions. It lies in wait just under the sand, with only its eyes and nostrils showing, and leaps out when the prey comes close.

Tropical rainforest – offshore island

MEGALODONTOSAUR

Megalodontosaurus modernus

The boughs and evergreen leaves of the tropical forest filter the sunlight into fine steamy beams that throw a dappled pattern down into the gloom of the jungle floor. This is the domain of the brightly coloured birds, the pterosaurs, with their larger wings, being banished to the more open spaces above. Suddenly the raucous chattering of the birds is hushed, and a deep still silence falls in the shadowy forest. The birds have sensed something dangerous. The light patterns ply upon the leathery surface of something dark and indistinct in the sparse undergrowth. With a gasp of air from cavernous lungs, a huge animal rises to its hind feet, pushing itself up on massive legs, steadying itself with clawed forefeet. The great eyes, gummy after sleep and torpor, blink and look around. The huge head contains only enough intelligence to tell the carcharodontosaur that it is hungry again. Slowly it crushes its way through the shadows to the decaying corpse of a plant-eater, scattering the scavenging birds and mammals, and resumes the meal that it abandoned days before.

Leathery skin? Low intelligence? It is as if time had been turned back to the Mesozoic period when the most important meat-eating dinosaurs were the vast lumbering carcharodontosaurs.

The huge island that the *Megalodontosaurus* inhabits lies off the east coast of the Ethiopian continent. Much of the geographical and climate changes that affected the rest of the world have passed this island by, and there has been little impetus for evolutionary change. This island is almost a time capsule, revealing what animal life was like on Gondwana hundreds of millions of years ago. Back then the biggest meat-eaters were the carcharodontosaurs, an offshoot from the allosaur group that were the top predators of late Jurassic and early Cretaceous times. Over most of Gondwana they were succeeded by a different line of theropods, the abelisaurs, in the late Cretaceous. Although the abelisaurs did get to the offshore island their tenure there was short lived. The carcharodontosaurs remained supreme, and culminated in the fearsome *Megalodontosaurus*.

▷ The classic shape of the large flesh-eating theropod dinosaurs – is retained in the **megalodontosaur**. Its sharp teeth and the powerful claws on the forelimbs are used for holding prey and tearing it up. The actual killing is done with great talons on the hind feet. As it walks the heavy head is held out at the front, balanced by the long heavy tailed behind.

◁ The skull of the modern **titanosaur** is almost identical to that of its ancestor of Jurassic and Cretaceous times. It has a steeply sloping face with nostrils at the top of the head and an array of peg-like teeth confined to the front of the mouth. The titanosaur does not chew but merely crops off its food and passes it down to be treated in its gizzard and its stomach.

△ The extremely long tail of the **titanosaur** is strong and muscular, and can be used as a whiplash to inflict a severe blow upon an enemy.

Dry open woodland – offshore island

TITANOSAUR

Altosaurus maximus

As the climate and vegetation on the offshore island changed little over the 145 million years since it became isolated, the animals that ate the vegetation did not need to change either. The large sauropod dinosaurs – the long-necked plant-eaters have remained practically unaltered over that period of time. The largest is the titanosaur that reaches a length of about 18 metres (60 ft) and can reach up to heights of 6 metres (20 ft) to browse from the tops of trees. The body is heavy and supported on stout pillar-like legs. The vertebrae of the neck, body and tail are partially hollowed out to cut down on the weight. The tail is usually carried clear of the ground and ends in a whiplash with which it can defend itself against enemies, such as the megalodontosaurus. Titanosaurs usually travel through the forests in large family groups, browsing constantly from the trees. The continual chewing of new shoots and buds results in trees on this island having naked trunks rising to a height of about 6 metres (20 ft) before the branches start growing. The food is not chewed but passed down into a voluminous gizzard. There it is ground up by the stones that are occasionally swallowed by the animal. From there the food passes into the stomach where it is broken down by bacterial action. The gizzard stones wear way quickly and the worn ones are often vomited up and replaced by fresh ones. The soil of the island is littered with small conical heaps of these rounded, discarded stones. There were many families of sauropod in Jurassic times but it is now only the titanosaurs that survive, and these are found in the Neotropical and Oriental realms as well as on the island – all fragments of the ancient continent of Gondwana.

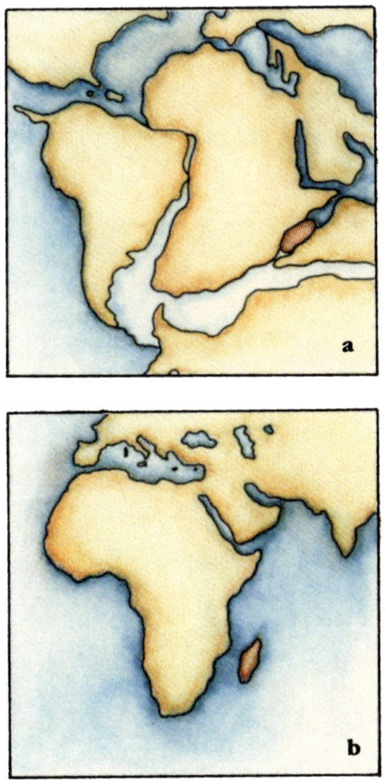

◁ Off the east coast of the Ethiopian continent, about 400 kilometres (248 m) out, lies a very large island. 1,300 kilometres (800 m) long, 600 kilometres (370 m) wide, it is a little continent and zoogeographic realm of its own. Since it broke away from Gondwana **(a)**, 145 million years ago, the island has remained relatively stationary and at the same latitude **(b)**. The original animal life has not evolved dramatically, as conditions there have been fairly constant, and much of the island is covered with the same forest that existed there in Jurassic times.

Shoreline – ocean islands

DWARF MEGALODONTOSAUR

Megalodontosaurus nanus

Black and white seabirds flock along the tidemark on the silvery sand, pecking at the seaweed and the dead animals washed up and left by the ocean. Other more adventurous birds strut about in the dry sand near the fringe of curving palm trees at the head of the beach. They are being watched from the ridge of red granite that sweeps down from the hilly interior of the island and protrudes as a headland at one end of the beach. Suddenly, from the rounded rocks, a sleek green and orange shape darts out onto the dazzling sand and, in a burst and flurry of feathers, one of the more foolhardy birds is caught and killed. The hunter is a tiny megalodontosaur similar to that found on the great island 1,000 kilometres (620 m) to the south. It is only about 3 metres (10 ft) long. It picks up its prey and carries it back to eat in peace in the darkness beneath the palms.

As Gondwana ripped itself apart and the continents were to become the Ethiopian and Oriental realms drifted away from one another, many fragments of continental materials were left scattered across the ocean in between. One particular fragment is crescent-shaped, about 1,000 kilometres (620 m) long and is almost totally submerged, but granite mountain peaks protrude as a scatter of islands about 1,000 kilometres (620 m) to the north-east of the island of megalodontosaurs and titanosaurs. Here again the animals are remnants of the old Gondwana fauna, but they are somewhat different, having evolved to cope with new conditions. On a small island there is not quite so much to eat, and the food is found in a much more limited area. Dwarf forms of animals evolve as the only types that are able to survive the more stringent conditions.

The dwarf megalodontosaur preys on dwarf versions of the plant-eating dinosaurs, but its build makes it agile and swift, and it can pursue fast-moving creatures, such as the seabirds that flock along the white palm-fringed beaches of the islands.

△ The **dwarf megalondontosaur** is a very agile and nimble animal. It usually hunts singly and can pray on the many seabirds that feed in flocks along the tideline. The megalondontosaur is athletic enough to sprint along the beach and snatch the birds out of the air as they flutter in panic into the sky.

△ The **dwarf titanosaur** is very like its large relatives. The main differences are in the slimness of the legs, the weak appearance of the neck, and the shortness of the tail. Like its relatives, it eats forest vegetation which it grinds up in its gizzard with the aid of stomach stones.

There are several islands in the group, and the four largest support different species of dwarf titanosaur. These differ from one another very slightly – a reflection of their diets.

a

b

◁ The dwarf species of the plant-eaters (**a**) are only about a fifth of the length of the full-sized forms. The dwarf meat-eaters (**b**) are relatively larger, being about one third of the size of the larger species. The meat-eaters are less restricted in the amount of food available as they also prey on sea birds.

Temperate forest – ocean islands

DWARF TITANOSAUR

Virgultasaurus minimus

The amount of plants available on an island leads to restrictions in the sizes of the plant-eating animals. Dwarf titanosaurs live among the coconut palms of the seashore and in the ferny undergrowth beneath the oaks and pines further inland. They feed on the undergrowth and the leaves of the low-growing vegetation. Unlike their larger relatives, the dwarf titanosaur is not a gregarious animal. It lives in small groups of two or three rather than in large herds. The group can move quite quickly through the thickets, long necks waving, when faced with an agile predator like the dwarf megalosaur. The body of the dwarf titanosaur needs smaller legs to support it, but the head is as large as that of a conventional titanosaur, and so it looks proportionally large. Dwarf animals are not unique to these islands. Dwarf forms of other large animals live on other islands in the oceans of the world, where a group of animals has become separated from the mainland population. It often happens that the isolated population has no meat-eaters among its numbers. Under these circumstances small and swiftly moving plant-eating animals may develop into large slow-moving forms, there being no need for speed to escape danger. On an island, therefore, there may be dwarf species of large animals, living beside giant species of small animals.

THE PALAEARCTIC REALM

The Palaearctic realm is the largest of the zoogeographical realms, covering as it does the largest continental area on the planet. From east to west the landmass stretches 17,000 kilometres (10,500 m), but from north to south its maximum width is about 7,000 kilometres (4,350 m).

Historically, the continent consists of the largest part of the supercontinent Laurasia (see page 16). On its western margin it is moving away from the other large piece of Laurasia – the Nearctic continent – but on its eastern point it remains very close to it. Many times in the past the continents have been united at this point and there has been a free interchange of animals across the isthmus that has formed there. The southern boundary has always been the Tethys Sea, but since Tertiary times other continents have moved northwards and united with it. This collision of continents has thrown up the greatest mountain chain in the world and this has acted as a natural barrier to migration, so defining the edge of the realm. In the south-west the continent that now holds the Ethiopian realm has not quite united with the Palaearctic realm. There is still an inland sea between them. The animal and plant life around this sea is broadly similar so this whole area is regarded as part of the Palaearctic realm as well.

Most of the mountains of the continent are quite recent and lie along the southern edge and around the inland seas in the south-western corner, although there are several older ranges. One of these fringes the islands and peninsulas of the north-west coast, and another runs north-south showing where two ancient continents fused together.

The habitats of the Palaearctic realm tend to run in parallel east-west bands. In the far north is the tundra, a treeless region of permanent chill enlivened by a brief summer. To the south of the tundra lies the largest uninterrupted forest belt on the planet. A vast expanse of coniferous trees stretches from one shore to another. This cold forest is home to a number of dinosaurs that are adapted to feeding on conifer needles and cones, and to the dinosaurs that hunt them. Further south again, the centre of the continent, far from the sea, is a dry grassland grading into desert towards the barrier mountains. This is the domain of hardy animals like armoured dinosaurs.

Closer to the sea, by the gulfs and archipelagoes of the eastern and western shores, the climate is less extreme, being modified by the regular moist winds from the sea. Deciduous forests grow here, and these support numerous varieties of dinosaurs, pterosaurs, birds and small mammals.

Barren land
Tundra and alpine
Coniferous forest
Deciduous and mixed woodland
Seasonal 'mediterranean' conditions
Steppe and grassland
Desert and desert scrub

Deciduous and mixed woodland

GESTALT

Formisaura delacasa

The most remarkable animal of the Palaearctic realm must be the gestalt. It is unique among the vertebrates because it pursues a communal existence. The gestalt evolved fairly recently, in the Ice Age, when there were few reptiles existing in the western portion of the continent and food was scarce. The animals that did survive there had to develop all kinds of strategies in order to make best use of the food available under the harsh circumstances. The gestalt evolved from one of the smaller pachycephalosaurs – the bone-headed dinosaurs – that lived in herds, mostly in mountainous areas.

The gestalt's strategy was to delegate the breeding of the herd to one individual female, freeing the rest to become food gatherers. They developed the ability to build citadels to guard their hard-won food against other animals. Within this citadel the individuals adhere to a strict discipline and a hierarchy that is based on age and sex. Now that the Ice Age has passed, the gestalt has found that its evolved way of life is still very efficient and successful.

Among the pachycephalosaurs the armoured head developed as a display structure, particularly in the males. In the gestalt, the male still has a very specialized head, but it is not used for fighting or intimidating members of its own species. The armour has developed spines containing a lethal poison, now used when protecting the colony from predators.

In times of plenty the population tends to grow too big for the colony to sustain. Small groups of adult males and females then leave to start up new colonies. Streams in the temperate regions of the Nearctic continent may be marked by lines of the conical citadels.

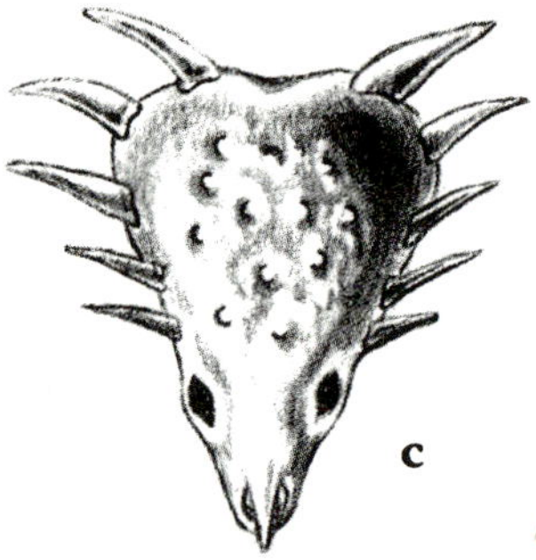

c

e

d

◁ The largest individual in the **gestalt** colony is the queen (**a**), almost a metre (3 ft) long and with a bloated body. She lays, on average, one egg per day all year round. Each egg hatches into a juvenile that is cared for by adult worker females that never leave the colony. When the juveniles, both male and female, reach an adolescent stage they work outside the nest. Adult females (**b**) eventually return to the nest and work as nursemaids. They are prevented from coming into a breeding condition by pheromones (chemical secretions) emitted by the queen. When the queen dies the pheromones stop and a new queen develops from one of the female workers. The female workers and the queen have very small eyes. They find their way about the gloom of the nest's interior by means of sensory hairs growing from their shoulders.

a

b

◁ The males are soldiers and have poison spines that grow outwards from the headshield (**c**). They stand guard at the nest entrances, warning of approaching danger by head-banging against the branches (**d**). Old males may come into breeding condition, shedding their spines (**e**), adopting a more subdued colour and living inside the nest with the queen. Breeding males are short-lived and are replaced every ten days or so.

◁ The nest is built and kept in repair by the females. It is a thatched structure, built of twigs and straw, usually around a sloping tree trunk over a stream. The interior is a mass of tunnels and chambers and each nest usually has the same layout. The egg chamber (**f**) is near the apex where it will be warmed by the sun. The queen chamber (**g**) is directly below. The nursery chamber (**h**), where the hatchlings are tended is below that. The toilet area (**i**) is directly over the stream. The food store (**j**) is attached to the main trunk and there are up to six additional food stores (**k**) on other branches of the tree. Long sticks and saplings are woven into the structure to supply escape routes if the main entrance along the tree trunk is rendered impassable by attack or weather damage.

▽ The food of the colony consists of buds in the spring, young shoots in summer and fruits and nuts in the autumn. The adolescents gather the food, passing it along a chain of individuals from one to another until it reaches the nest. This chain is guarded on both sides by adult males.

Deciduous and mixed woodland

BRICKET

Rubusaurus petasus

The deciduous woodlands of the Palaearctic realm are found mostly north of the mountains at the western end of the continent, where the continent narrows and few places are far from the sea. It is a region of high rainfall and temperate climate. There are four marked seasons: dormant winter; spring in which leaves and flowers appear; summer with the most vigorous growth; and autumn, which is a time of fruiting just before the trees lose their leaves for winter. The predominant trees are oaks, ashes and beech, below which is usually an understorey of smaller trees and a thick under-growth.

A typical animal of this environment is the bricket, a small browsing hadrosaur, not dissimilar from its Cretaceous ancestors. In the Cretaceous period there were only the non-crested hadrosaurs living in this corner of the continent, but later the crested forms migrated here from further east - part of the great spread of the hadrosaurs over the northern continents. The bricket lives in small herds in the dense undergrowth and bramble thickets, usually resting during the day and feeding at dusk and at dawn. The expanded crest, found in both males and females, is both used as a display structure, particularly during the autumnal mating season, and as a deflecting device when it must move swiftly through the vegetation.

▽ The long flat tail of the **bricket** is used both as a prop **(a)**, when browsing from high branches, and as a warning flag **(b)** at times of danger. Stuck straight up in the air its bright colours warn the rest of the herd of approaching predators.

◁ Ticks, fleas and other parasites are easily picked up by thicket-living animals. The **bricket** has a cleansing ritual that deals with this. A bricket suffering badly from parasites seeks out a piece of fur or a mat of hair that has been lost by another animal on the bramble thorns. Then, holding the fur in its beak **(1)**, it walks backwards into a river **(2)**, very slowly, until it is totally submerged, but for its snout. The parasites move up the body, the neck and the head, and are eventually all stranded in the piece of fur. The bricket then abandons the fur and its passengers to the current **(3)**. There usually follows a period of playful high spirits in the water as several newly cleansed brickets frolic with one another. Mating usually takes place at this time. Further downstream the fur is seized by hungry zwims that feast on the parasites **(4)**.

Deciduous and mixed woodland

ZWIM

Naremys platycaudus

In outward appearance many of the mammals have changed little since they evolved from the paramammals in Triassic times. Throughout the Jurassic, Cretaceous and Tertiary periods they have remained small, compact creatures, not adapting into any of the wide ranges of lifestyles occupied by the great reptiles. However, some of them have a number of interesting specializations.

The zwim is an aquatic, insect-eating mammal. It inhabits the streams and rivers of the Palaearctic realm and is particularly common in regions of deciduous forest. It has a length of about 30 centimetres (1 ft), most of which is taken up by a long flattened tail. The tail, and the long webbed hind feet, allow the animal complete freedom in the water. Its long sensitive snout is used for probing under stones and in dead vegetation for the insects and other invertebrates on which it feeds, both at the bottom of the stream and on land. It lives in burrows on the riverbanks and can defend itself against predatory reptiles and fish by biting with its sharp teeth. The saliva is venomous and any bite is quickly effective. The zwim is a social animal and as many as a dozen burrows can be found within a short distance from one another on heavily wooded riverbanks. Large numbers may congregate at the wallowing-pools of the bricket in order to feast on the parasites that are shed there.

◁ The **zwim** is an active swimmer, thanks to its webbed hind feet and its flattened tail that works with a strong up and down undulation. The eye is large and it can adjust its focus to see both underwater and on land.

◁ The **bricket's** streamlined shape is ideal for fast movement through the tangles and thickets of the temperate woodlands. The shape of the crest **(c)** parts the vegetation as the animal runs, and the slim body allows it to pass between close-growing trees. The brownish colour camouflages the animal when motionless, but when it breaks cover and runs it can do so quickly, vanishing at speed into the depths of the forest.

The **coneater** is an animal of small herds, each herd consisting of about a dozen individuals. Its body is insulated from the intense winter cold by deep folds and wrinkles of fat. Its beak can snip twigs and cones (**a**), and its tough food is ground up between batteries of grinding teeth at the back of the mouth.

a

Coniferous forest

CONEATER

Strobofagus borealis

The coniferous forest of the Palaearctic realms is the largest area of unbroken forest on Earth. It stretches right across the continent along the sub-polar latitudes and, but for the narrow inlet of the sea, would be continuous with that of the northern Nearctic realm. It is bounded by the tundra to the north and the deciduous forests and cold grasslands to the south. The coniferous forests contain relatively few species, compared with other forests further south. The most common conifers are pines, firs, spruces and larches. These reproduce by means of seed-dispersing cones, and it is these cones that provide the food for most of the animal life of the region. Almost the only large animal found here is the coneater, a 3-metre-long (10 ft) hypsilophodont that resembles its Cretaceous ancestors.

The hypsilophodonts were a very widespread and successful group in Cretaceous times. Lightly built running animals, very much like small iguanodonts, they spread over all the continents before the end of the period. They have continued to be successful all over the world until the present day, occupying a large number of different niches. In the dark depths of the coniferous forest they run in small herds, over the soft undergrowth-free forest floor, and browse in more open country along the river banks. Although the coneater generally eats the cones and the seeds they contain, in winter it eats tree bark, needles, mosses and lichens, and seeks out stores of nuts hoarded by smaller animals.

◁ The **jinx** usually hunts in pairs. A pair insinuate themselves into a herd of coneaters as they move through the forest (**1**). The jinx's scent glands give off a smell similar to that of its quarry, aiding the disguise. Suddenly, when the herd least expects it, one jinx kills a coneater. The rest of the herd, including the other jinx, scatters (**2**). When they come together again after the first panic passes (**3**), the second jinx attacks and also kills.

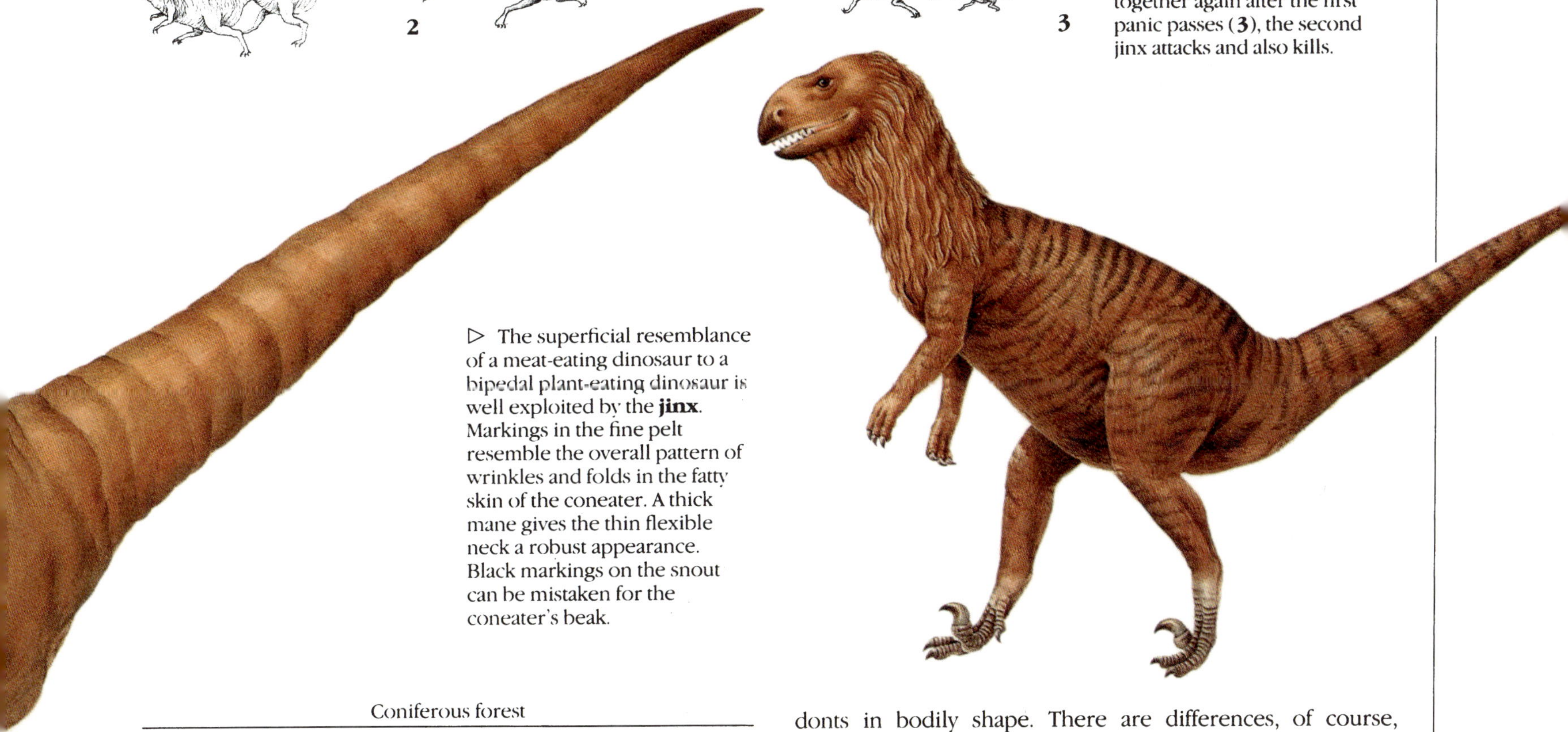

▷ The superficial resemblance of a meat-eating dinosaur to a bipedal plant-eating dinosaur is well exploited by the **jinx**. Markings in the fine pelt resemble the overall pattern of wrinkles and folds in the fatty skin of the coneater. A thick mane gives the thin flexible neck a robust appearance. Black markings on the snout can be mistaken for the coneater's beak.

Coniferous forest

JINX

Insinuosaurus strobofagoforme

There is not much to eat in the coniferous forest compared with the food available in other parts of the world. Hence there are fewer types of animals here, and each type is very highly adapted for its particular life style. Nowhere is this more evident than in the case of the jinx. The jinx eats coneaters, and nothing else. Its whole physiology has developed to aid this. Since the Cretaceous period, the most successful medium-sized carnivorous dinosaurs have been the dromaeosaurs and the saurornithoids. These are generally lightly built bipedal animals, about 3 metres long (10 ft), with long, stiff tails, which means they resemble hypsilophodonts in bodily shape. There are differences, of course, especially in the teeth and claws. Both the saurornithoids and dromaeosaurs have a large killing claw on the second toe, obliging them to walk on only their third and fourth toes. This is quite different from the three-toed arrangement of the feet of the hypsilophodonts. The long mouth conceals a battery of meat-shearing teeth. They have grasping hands but with only three fingers, compared with the five of the plant-eaters. The meat-eaters also lack the characteristic pot belly of the browsers. Nevertheless, the superficial resemblance is so close that one dromaeosaur, the jinx, has adopted a coloration that disguises it as a hypsilophodont and allows it to infiltrate the coneater herds. The deception is so successful that a herd of coneaters may travel for quite some distance without noticing the danger in its midst.

◁ The **tromble**, a 3 metre (10 ft) high flightless bird, with legs as massive as tree-trunks, migrates in huge herds across the waterlogged landscape of the summer tundra *above*. Eggs are laid in temporary nests at the northernmost point of the migration, and they hatch very quickly with the young able to travel immediately.

Tundra and alpine

TROMBLE

Gravornis borealis

△ The very large body of the **tromble** helps to retain its natural warmth. A large animal such as this has a small surface area compared to its bulk, and heat cannot escape easily. The thick coat of hair-like feathers helps the insulation, especially during the winters spent in the shelter of the coniferous forests to the south.

To the north of the vast belt of coniferous forest that circles the northern hemisphere in the bleaker sections of the Palaearctic and Nearctic continents, the trees give way to a cold region of open landscapes called the tundra. Here the conditions are too tough even for the hardiest of trees, and the vegetation consists of short grasses, mosses and lichens. During the long northern winter, when the sun does not rise for weeks or months at a time, nothing grows. Then, when the bleary summer finally dawns, the winter snows melt away and all the plants grow frantically to best harness the short living season. Animals migrate northwards from the coniferous forests where they have wintered, to take advantage of the sudden harvest of food. They spend the summer months moving northwards, grazing as they go.

The largest animal to migrate in this fashion is the tromble, a massive flightless bird that evolved when the tundra regions appeared during the Ice Age. There are no dinosaurs this far north. Even though the dinosaurs developed a warm-blooded system that enabled them to survive a great range of climates, the chill tundra environment was still too extreme for them. The warm-blooded physiology of birds was much more efficient and therefore the birds were able to move into this harsh ecosystem. Several ground-dwelling birds evolved here since there were no terrestrial predators to threaten them.

△ The beak of the **tromble** is broad and hard, for cropping great mouthfuls of the coarse tundra vegetation. The food is ground up in the crop with the aid of swallowed stones. Fresh stones are available every year, brought to the tundra surface by frost. The mating season is early summer, when the males sprout bright yellow display plumes.

Tundra and alpine

WHIFFLE

Adescator rotundus

▽ Several species of **whiffle** are found on the tundra, but they are all rather similar to one another. The body is round and the neck short, to minimize heat loss. Long slender legs enable it to wade in ponds, and the long beak, with the sensitive tip, can probe into mosses and under stones for insects.

Summer in the tundra reveals not only a rapid growth of plants, but a sudden flourishing of insects as well. Thick black clouds of flying creatures rise from the waters and hover like smoke over the tundra landscape. The mosses and lichens are, for a few short months, alive with scurrying beetles, springtails and mites. Such a harvest of food cannot fail to bring in flocks of insect-eating birds, and huge formations can be seen swooping and diving above the lakes and bogs in the summer.

One particular bird has abandoned its powers of flight, leaving the clouds of flying mosquitoes, midges and caddis flies for the airborne hunters, and has concentrated its feeding to the insect life on the ground. Large ground-scurrying flocks of whiffles follow the tromble herds, pecking here and there in the stunted vegetation with their long beaks. Through the weight of their passage, the trombles cause such a disturbance to the soil and moss, churning up the mud and crushing down the vegetation in deep foot-prints, that the insects that live there are sent scattering for new cover. The whiffles following behind are quick to snap them up. Each tromble, representing a huge mass of flesh, is prey to warble flies, fleas and all manner of other parasites which are also eaten by the whiffle flocks.

▽ When asleep, the **whiffle** can draw its head into the hairy plumage of its body, and hide its sensitive beak among the narrow feathers of its chest. Like most tundra creatures, the tromble and the whiffle are found in the cold northern regions of both the Palaearctic and Nearctic realms.

Steppe and grassland

TARANTER

Herbasaurus armatus

In the windswept heartland of the Palaearctic continent, the continuous belt of coniferous forest is replaced to the south by cold grassland of mixed grasses and narrow-leaved flowering plants. At the forest edge the grasses are rooted in a rich black soil, but further south the soil is redder and drier. These grasslands are known as the steppe. Beyond the steppe a wilderness of drifting sand, hard-packed clay and shattered stone sweeps southwards to the foothills of the huge mountain chain that marks the southern limit of the Palaearctic realm.

The steppe grassland is home to many small seed-eating burrowing animals, particularly mammals. Their burrowing way of life keeps the soil churned up and aerated and prevents all the rich topsoil of the surface layers from being used up too quickly. In this way the plant-life is maintained. In the spring the steppe is a blaze of colour as the flowering plants come into bloom, while in summer the feathery seedheads of the grasses create a totally different landscape.

As in most of the great grasslands of the world, the steppe supports a variety of grazing animals, all of which have evolved since the Mesozoic era. The taranter has evolved from the ankylosaurs, such as *Euoplocephalus* and *Saichania* that became so plentiful at the end of the Cretaceous period. The ankylosaurs were divided into two groups. The first were quite lightly built and came to their peak early in Cretaceous times. The more advanced group were heavily built, with massive armour and a weapon on the end of the tail, and were successful after the first group began to decline. Most of the ankylosaurs that exist today are descended from the heavily armoured forms. They were always abundant in this part of the world and fossils reveal that they evolved into grazing animals as the grasslands developed. In the taranter the armour, developed from horn-covered bones set in the skin, has become a continuous covering for the back. This, and the bulbous shape of the body, help to prevent desiccation in the dry winds.

▷ Ankylosaur armour, consisting of horn-covered bone, evolved as a defensive mechanism. In the **taranter** it has become more important as a means of conserving moisture. Hollows inside the skull are lined with damp membranes that moisten the dusty air as it is breathed in. Defensive armour is still present as horny spikes along the flanks and the heavy club on the tail.

△ The head armour of the **taranter** is a horny covering that forms a grass-cutting beak along the edges of the broad mouth. Flat teeth in the back of the mouth grind up the plant food.

Desert and desert scrub

DEBARIL

Harenacurrerus velocipes

The deserts of the southern edge of the Palaearctic realm are among the harshest environments on Earth. Vast expanses of sand, clay, rock and rubble are baked by a broiling sun during the day and frozen by cold at night. There are no barriers to the chilling winds that blow southwards from the northern icecap, or downwards from the mountain peaks in the south. Rainfall is infrequent and confined to spring and autumn. The vegetation is sparse and only appears when rain has fallen. Animals must be quite hardy to exist in such an environment, and most of the plant-eating creatures found here are cold-blooded. The slow metabolism of such herbivorous animals as tortoises and plant-eating lizards enables them to gorge themselves when the plants are available, and to sleep away the harsher times in burrows. Of the warm-blooded animals the mammals are most successful, avoiding the extreme conditions by being active only at dawn and dusk.

One of the few warm-blooded dinosaurs that live here is the diminutive debaril. In its life style and, to some extent, its appearance, it parallels the desert mammals. It is active at dawn and dusk, eats roots and seeds, conserves its water efficiently and burrows in the sand.

▷ The **debaril**, like most of the small, plant-eating modern dinosaurs, is descended from a hypsilophodont ornithopod. It is about 60 centimetres (2 ft) long and is well adapted to the extremes of its environment. Under cold conditions it hunches itself up to present a small surface area to the wind (**a**). The wrinkles in the skin bunch together to form an insulating coat. Heavy lids fold over the eye to prevent chilling. During cold conditions it moves in a series of bounds (**b**). When it is hot it stretches itself out (**c**), presenting a large surface area that is easily cooled, and travels by running on hind legs (**d**).

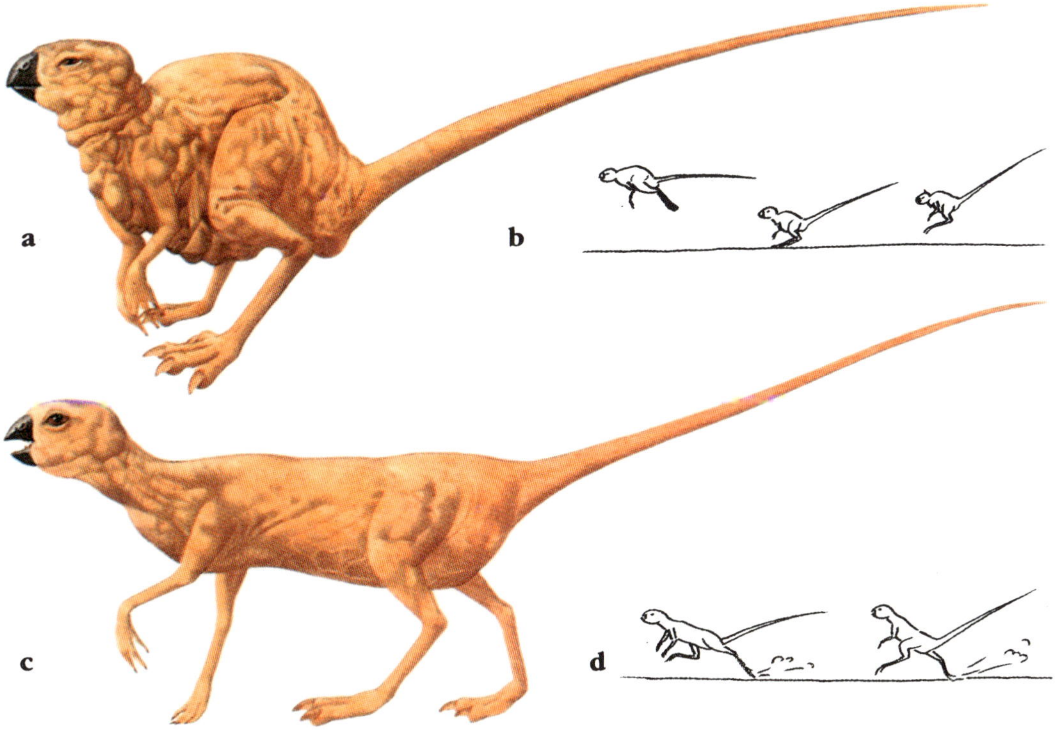

a

b

c

d

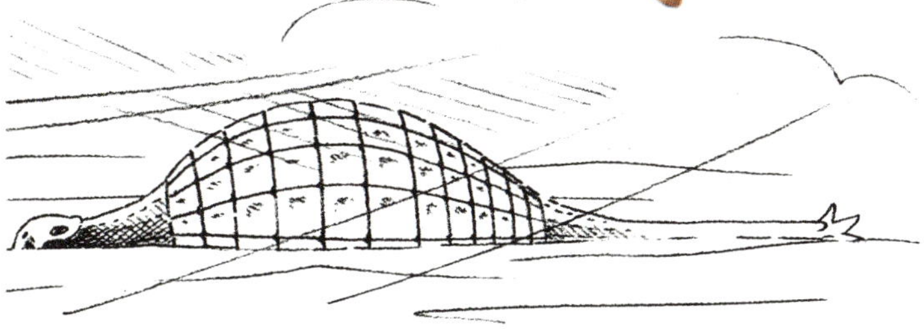

▷ By settling down into a hollow and presenting its streamlined armoured surfaces to the wind the **taranter** can withstand the stinging sand and duststorms that are common in the area.

THE NEARCTIC REALM

The Nearctic realm consists of a roughly triangular complex of continent and islands, reaching from the desert latitudes in the south to the polar ice cap in the north. At its broadest it runs along the Arctic Circle where it consists of glaciered mountains and icy islands in a frozen sea. To the south, in its more habitable zone, it is a continent 5,000 kilometres (3,000 m) broad, but this narrows rapidly to an isthmus connecting it to the Neotropical realm in the south. Much of this isthmus is desert, and provides the barrier between this realm and its Neotropical neighbour.

The continent once occupied the western part of the supercontinent of Laurasia. It was in this region that the two supercontinents, Laurasia and Gondwana, were almost permanently united in the days of the world continent of Pangaea. It did not split away from the continents that were to become the Ethiopian and the Neotropical realms until late Jurassic times. At about this time, too, it broke free from the remainder of Laurasia, as a new ocean opened up between it and the rest of the supercontinent to the east. Since then, land connections with other continents have come and gone many times. In the north-west the most westerly portion is separated from the most easterly point of the Palaearctic realm by the shallowest and narrowest of seaways. Throughout its history this seaway has dried out and flooded again, each time allowing animals to cross from one continent to another. At the moment it is flooded and forms the boundary to the realm, but that will only be a temporary state. Likewise, to the south, the isthmus between the Nearctic and Neotropical continents is a very temporary affair, breaking and re-uniting many times in the past, and each time having an effect on the migration of animals. At the moment the connection is complete, but that may not be for long.

Topographically the continent consists of new mountains in the west, old mountains in the east, and flat plains between. The western mountains are still building up, due to the movement of the oceanic plate there, and earthquakes and volcanoes are frequent. The whole of the northern portion of the continent, much of it submerged and only appearing as islands, consists of hard, old Precambrian rocks, the kind that form the nuclei of all continents. The southern half of the lowland section consists of a vast river basin draining to a gulf in the south.

The habitats range from frigid cold to searing hot climates. The far north is still in the grip of the Ice Age, with continental glaciers smothering the great north-eastern island, and pack ice choking the straits and inlets. Where the land is not ice covered it consists of tundra - bleak and only habitable for a few short months of the year. Even if the land bridge between the Nearctic and Palaearctic realms were not currently broken, these conditions would prove just as impenetrable a barrier between the two areas.

Further south lies a vast belt of coniferous forest, almost a continuation of that region found in the Palaearctic realm. The animals there are similar to those in the Palaearctic forests, as there was a clear connection between the two realms until very recently.

Deciduous forests lie to the south of this and clothe the older mountain ranges found in the east. Here, the tree-living animals, particularly the arbrosaurs, are superficially similar to those that live in deciduous forests elsewhere. They have been separated from them for so long, however, that these arbrosaurs are of totally different species.

The western third of the continent consists of mountain ranges - high and elongated, with desert valleys and plains between them. Both plant-eating and meat-eating dinosaurs have adapted to the sparse food supplies in the high reaches.

The central lowlands, drained by the vast river system, is a region of temperate grasses. The grass is long and lush by the rivers but becomes shorter and sparser as the land dries out and rises towards the western mountains. Hadrosaurs; evolved originally to live in lush forest, now roam these open prairies, having adapted successfully to a grazing existence. With the development of fleet-footed grazing animals, comes the evolution of carnivorous hunting animals.

Low-lying lands around the south-eastern corner of the continent, where the land has only recently risen from the sea, and the vast deltas produced as the huge river empties into the gulf to the south, have given rise to a unique series of water-living animals. Here are dinosaurs that hunt fish, and pterosaurs that strain tiny invertebrates from the shallow water. There is even a hunting dinosaur that demonstrates a cunning intelligence unparalleled in the dinosaur world, to enable it to catch swift flying creatures such as birds and pterosaurs.

Barren land
Tundra and alpine
Coniferous forest
Deciduous and mixed woodland
Seasonal 'mediterranean' conditions
Prairie and grassland
Desert and desert scrub

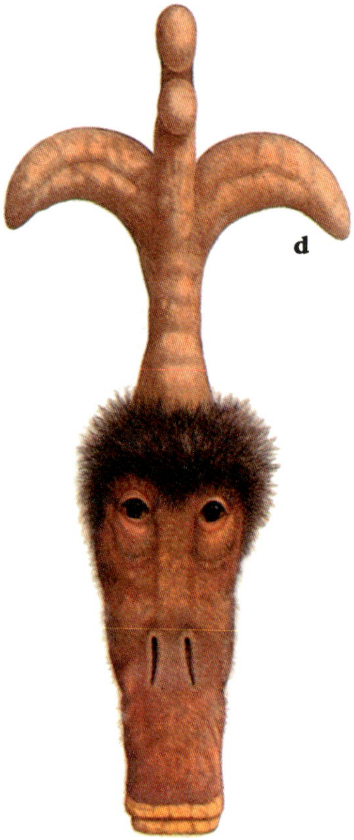

d

f

e

Crests vary from species to species and are used as signalling and identification devices. The shapes range from hook shapes, as in *Ancorachephalus major*, (**d**), through rows of knobs, as in *Sprintosaurus quadribullus*, (**e**), to broad blade shapes, as in *S. dolabratops*, (**f**).

◁ The **sprintosaurs** evolved from the duckbilled dinosaurs of the Cretaceous period. These, in turn, evolved from completely bipedal dinosaurs (**a**) but began to spend more time on all four legs (**b**). When they evolved into the plains-living sprintosaurs, they became completely quadrupedal (**c**). The legs are long and lightweight, with muscles concentrated in the thighs and upper arms, and the tail, having lost its balancing function, has degenerated either to a stump or to a flagpole.

a

b

c

Prairies and grassland

SPRINTOSAURS

Family Sprintosauridae

The prairies – the grasslands that spread out across the centre of the Nearctic continent from the great winding river on the lowland plain to the towering wall of the massive western mountains – support vast herds of grazing animals. The most widespread of these are the sprintosaurs, evolved from the duckbilled hadrosaurs of the late Cretaceous. In the last 65 million years, as the forests gradually gave way to the grasslands, the hadrosaurs evolved to adapt to the changing habitats. Like the hadrosaurs before them, there are two main groups of sprintosaur, the crested, and the non-crested.

The **crested sprintosaurs** tend to inhabit the western high prairies, grazing the short grasses and undergrowth of prickly pear cactus in the dry shadow of the mountains. In the long evenings of the mating season the thin upland air resonates to the shrill trumpeting sounds as the sprintosaurs call to one another, using the hollow crests as sounding tubes. The crests are formed from the bones of the nose. They probably evolved to extend the nasal passage and allow dry air to pass over moist membranes before reaching the lungs. They successfully perform this function, because the air of the high prairie is very dry. Distinctively shaped crests are found in different species of crested sprintosaur and serve to distinguish members of one species from another, as they roam the sparse grasslands below the mountain ramparts. Both male and female sprintosaurs possess the crest.

The **non-crested sprintosaurs**, like their slightly larger crested cousins, evolved from the hadrosaurs. However, while the crested forms evolved from such crested hadrosaurs as *Corythosaurus* and *Parasaurolophus*, the non-crested forms evolved from the flat-headed line of the hadrosaur group that included *Hadrosaurus* and *Anatosaurus*. Since Cretaceous times, their evolution into a grassland animal was parallel to that of the crested sprintosaurs. The cropping beak and the huge battery of constantly replaced grinding teeth needed little modification to suit a diet of tough, silica-rich grasses. As their faces were permanently in the grass, eating, the animal adopted a four-footed posture. On the open plains danger can be seen from a long way off and so the legs became spindly and lightweight to enable the animal to run swiftly from harm. The face became long to keep the eyes above the level of the grass so that a lookout could be kept for predators while grazing.

The main difference between the non-crested and crested sprintosaurs, apart from the crest, is the presence of the tail. Having lost its function as a counterbalance, necessary when its ancestors went about on two legs, the tail has become a tall, stiff flagpost. The males have a fin of skin supported by bony struts at the tip, and this is used for signalling. Different species have different sizes and patterns of fin. The non-crested sprintosaurs live mainly on the long-grass, lowland prairie where the adaptations of the particularly long face and the tail flag are of most use. The sprintosaurs, both crested and non-crested types, travel over the plains in closely knit herds.

Vexillosaurus levipes is a **non-crested sprintosaur** from the floodplains of the great central river. It moves about in tight herds often only seen as a bunch of tails waving above the long grasses. When a predator such as the northclaw *Monuncus* attacks, the herd breaks up in a confusing flurry of flags and poles, leaving the attacker bewildered.

◁ The **northclaw** is a therizinosaur. These were a group of heavy slow-moving theropods that evolved in Cretaceous times. They were omnivorous and carried huge claws on their hands. Despite their clumsy appearance they were closely related to birds. Their descendants have now reverted to their original bird-like condition and are now completely carnivorous.

Prairie and grassland

NORTHCLAW

Monuncus cursus

Puma-like, the northclaw slinks through the grasses of the prairie, its tawny stripes blending it into the dry yellows and browns of the vegetation. Its reptilian eye is fixed on an unconcerned grazing group of crested sprintosaurs, its instinctive cunning guiding it towards the most effective attack. For long hours its sleek body moves, slowly but surely towards the unsuspecting herd. Suddenly a male sprintosaur, looking round, notes something in the grass that does not quite fit. He trumpets out a warning blast and the herd scatters in a cloud of dust. With an instant reflex the northclaw darts up from its place of hiding and arrows down on the slowest of its prey. Its powerful hind legs thrust the horizontal body forward, balanced by the stiff rod of a tail. Then, when almost upon the young sprintosaur, which is still deciding which way to run, the single long claw springs out, hooks into the skin and pulls the animal over, kicking and struggling in the dust and the grass. Another blow of the claw and the prey is disembowelled, and the northclaw settles down to feast.

▷ The horn of the **monocorn** is a formidable weapon when turned on an attacking carnivore like a northclaw. Monocorn herds usually travel with big males on the outside protecting the females and hornless young.

Prairie and grassland

MONOCORN

Monocornus occidentalis

The lightly built sprintosaurs are not the only grazing animals that roam across the vast prairies. Huge slowly-moving grass-eaters also exist here, cropping the grasses down to their roots and moving on in black dusty herds. The huge monocorn is one of several species of ceratopsian that still inhabit the Nearctic continent. It differs in appearance from the members of its ancestral stock, such as *Triceratops* and *Styracosaurus*, but the differences are not really profound, and they are a reflection of the life that the animal now leads. The herds of monocorn need to be on the move constantly, for once all the grass in one area is eaten up they have to move on to fresh areas. The legs are therefore longer and more slender then we would expect in such a large animal. The feet are digitigrade, that is they support the weight of the body on the toes rather than on the flat of the foot – the plantigrade condition of the monocorn's ancestors. In common with the ancient ceratopsians, the neck and shoulders are protected by a bony frill, and a horn on the nose is used as a weapon.

The ceratopsians are now no longer confined to the Nearctic continent. Several species are now found in the Palaearctic realm where they spread via the land bridge between the Nearctic and Palaearctic continents before the Ice Age.

△ The bony frill of the **monocorn** is very long, covering the length of the neck to the high shoulders. It is used when males spar with one another for herd leadership. The contestants push harmlessly against one another until one tires and gives way.

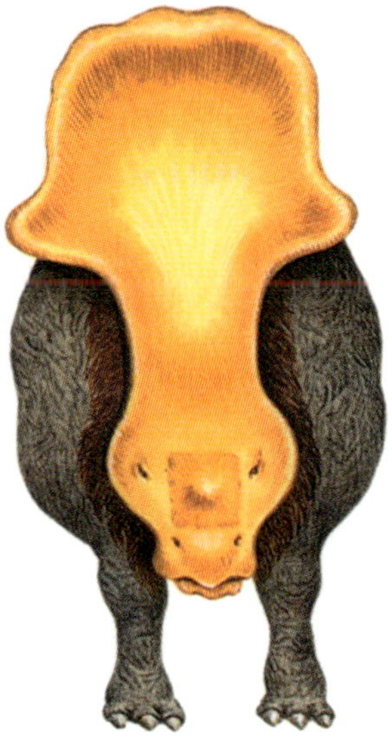

△ The **balaclav** lives in small family groups in the highest mountains. It can often be seen trekking across snowfields and glaciers from one lichenous rock or mossy hollow to another. It can subsist on a very poor diet.

△ Layers of fat insulate the **balaclav**, and the hairs on the tail and feet help it to grip icy surfaces. Its broad beak and the spade-like nails on the three middle fingers enable it to scrape up lichens and moss, while the longer claws are used for digging up alpine plants.

Tundra and alpine

BALACLAV

Nivesaurus yetiforme

The tangle of braided mountain chains in the western third of the continent is relatively new, thrust up to its present height only in Cretaceous times. The open river plains of the Jurassic and the steamy forests of the Cretaceous period – the habitats of the most prolific dinosaur communities in the past – are now craggy mountains, clothed in glaciers and snow. A few dinosaurs adapted to this new environment, moving in from the regions round about and adopting a lifestyle that would support them there. One such is the 2 metre (6½ ft) long balaclav, one of the specialized modern hypsilophodonts. It evolved insulating layers of fat and fur, an ability to eat the alpine plants and mosses, and a compact shape to preserve its body heat.

Tundra and alpine

MOUNTAIN LEAPER

Montanus saltus

The mountain leaper is evolved from the alvarezsaurs – a group of theropods with a single claw on each hand. The large brain of the mountain leaper coordinates its movements, and enables it to make swift judgements as it springs and leaps between the spires and crags of the mountain peaks, hunting the birds and small mountain mammals on which it feeds. It has developed a high degree of endothermy – warm-bloodedness – to help it to cope with the icy climates, and the thick shiny fur is indicative of this, providing the insulation that keeps the animal's body at a constant temperature. The head and body reach a length of about 1 metre (3 ft), but the flowing hair, the long legs and long balancing tail with its spectacular plume make it look much larger. Mountain leapers live in small packs, with the males undertaking all the hunting, and protection of the females and the young.

▽ When not hunting the male **mountain leapers** stand as motionless sentinels, guarding the females and the young of the pack. On bright days the pack may sun itself on the open slopes, at which time it is vulnerable to attack by birds of prey and mountain pterosaurs.

◁ **Mountain leapers** are agile and swift when traversing the mountain peaks and crags. With surefooted springs and leaps they move rapidly, maintaining balance with the long tail.

Mixed woodland – wetlands

SPRINGE

Necrosimulacrum avilaqueum

In the midst of the southern deltas of the Nearctic continent there lives one of the most cunning of the dinosaur hunters. Descended from the maniraptorans, such as *Stenonychosaurus* and *Saurorinithoides*, the springe, measuring about 3 metres (10 ft) long, has a similar size and proportion to its ancestors. Its head, however, has expanded and now contains a larger brain, and the killing claw is carried on a particularly long second toe. With its naked skin mottled a deathly white and pink, and the matted, dark patterned fur, the springe has a derelict, morbid appearance.

▷ The springe is the most cunning of the hunting dinosaurs. Its Cretaceous ancestors were the most intelligent animals of the time, and this trait has continued through evolution. The intelligence has evolved for a simple purpose – to find food more efficiently.

2

1

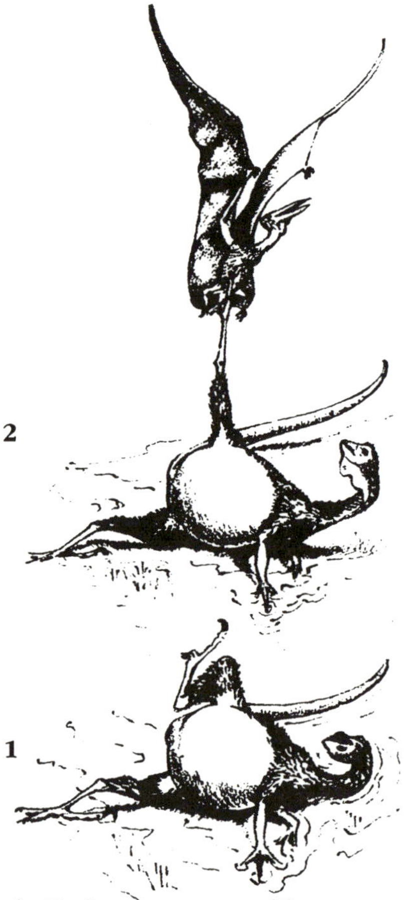

△ The hunting strategy of the **springe** is one of ambush. It lies on a tidal mudbank, in an attitude of rigor mortis, with its head and tail thrown back and its hind leg pulled into a stiff pose. It inflates its belly, showing off the death-like mottling, and emits a smell like that of putrefaction (**1**). This performance is irresistible to the carrion birds and scavenging pterosaurs of the swamps who flock to the site. A swift dart of the killing claw (**2**) and a victim is impaled.

▷ The open skies above the swamps and lakes are usually filled with wheeling flocks of birds and pterosaurs, particularly at dawn and at dusk in the evening. At a distance, pterosaurs like the **sift** can be distinguished from birds in flight by their comparatively larger wings.

Mixed woodland – wetlands

SIFT

Pterocolum rubicundum

A great tract of land on the southern and south-western shore of the Nearctic continent consists of deltas, swamps and backwaters. Since Cretaceous times it has gradually been rising above sea level, and it is now covered in marshy vegetation, comprising vast areas of reed beds and mangroves with stands of conifers, evergreen oaks and magnolias on the drier banks. The available moisture and warm climate mean that several varieties of plants and animals live here. Bird and pterosaur life is particularly abundant, as the birds have adapted well to this pleasant environment. Ducks graze the water weed, divers plunge into the waters after fish, and waders peck after small creatures in the mud and shallows. However, the pterosaurs live side-by-side with the various species of birds and reveal similar adaptations to the same conditions.

The sift is built to the general pterosaur design, with a small lightweight body stiffened for flight, wings of membrane supported by the arms and extended fourth finger, and a very flexible neck. However, in appearance the sift is more like a wading bird, with its long-shanked legs and its tapering beak-like jaws. The sift congregates in flocks out in the shallows. The folded wings, larger in proportion than those of birds, catch the sun and provide an everchanging pattern of light as the flock moves about filtering the tiny insects and crustaceans from the muddy water. The long, thin jaws are armed with a multitude of tiny, comb-like teeth which are used to trap the water-borne food.

△ The **sift** feeds on the tiny plants and animals that abound in the shallow waters of the deltas. Shrimps, worms and little fish are stirred up by the long-toed feet and caught with the narrow jaws. Floating algae and weeds are strained from the water by the fine teeth.

△ The second finger of the **nauger** is remarkable, being about as long as the forearm. It is used for poking down burrows and winkling out the larvae of the wood beetles on which the arbrosaur feeds. A hooked claw at the end secures the catch.

Deciduous and mixed woodland

NAUGER

Picusaurus terebradens

In the deciduous forests of the Nearctic realm live a vast variety of different arbrosaurs, each one specifically adapted to a particular way of life. One of the most highly specialized is the nauger, with its wood-boring jaws and its long thin finger. It feeds almost exclusively on the grubs of wood-burrowing beetles that it hunts in the living wood of the trunks and branches. The strong hind legs and the stiff bristles on the tail give it a firm grasp on the tree while it listens for movement beneath the bark and drills into the wood after the larvae.

If there were no dinosaurs living, it is possible that some of the ecological niches now occupied by tree-living arbrosaurs would have been occupied by birds. It seems entirely probable that a bird could have evolved to fill the niche of the pecking arbrosaur, with a strong bill taking the place of the powerful teeth, and possibly a specialized tongue doing the work of the long finger.

Deciduous and mixed woodland

TREEPOUNCE

Raminsidius jacksoni

As on all the other continents, the arbrosaurs fill the trees of the Nearctic realm. Some eat insects, others consume fruit and berries, and some prey upon other arbrosaurs. The treepounce is one such carnivore. With a head and body reaching 70 centimetres (2 ft) it is larger than most of the arbrosaurs, and so it tends to be less agile. However, what it lacks in agility it makes up for in stealth. Lying along a branch on a sunny day its spotted coat makes it almost invisible in dappled sunlight. With infinite patience it waits until an unsuspecting nauger, or smaller arbrosaur scrambles close, and then pounces, making a quick kill.

▷ The skull of the **nauger** is perfectly adapted to its purpose of drilling into solid wood. The teeth grow only at the front of the jaw and are directed forward, each one lending support to the one before. Those at the very front bear the brunt of the pecking force. When they wear out or break, they are replaced by more teeth growing in from behind. The neck joint is very strong, protecting the back of the skull and the brain and giving support to the heavy neck muscles needed for the constant rapid pecking.

▽ The **treepounce** relies mainly on its hearing for hunting, and so its ears are highly developed. Tufts of hairs around the ear-opening funnel the sounds into the ear canal. In appearance the tufts resemble the fleshy external ears found on some smaller mammals. The colouring of the coat conceals the treepounce in the trees.

Deciduous and mixed woodland

FOOTLE

Currerus elegans

The trees are alive with myriads of tiny insectivorous arbrosaurs, each one differing from the next by the different colours of the pelt, by different display tufts on the head and tail, and by slight differences in the shape and size of the head. The arrangement of the skull and jaws depends upon the diet. Feeding both in the trees and on the ground, the tiny arbrosaurs have short, thick jaws to crunch up beetles, or long thin jaws to dig for buried larvae and worms. All the arbrosaurs have the same light build, long, springing legs and thin toes. The body is small and balanced for running and jumping by the long, stiff tail.

The footle is a typical narrow-jawed arbrosaur. It is about 50 centimetres (1½ ft) long but most of this is made up of the long tufted tail. The body weighs only a few grams. It is a very agile little creature, scampering along boughs and leaping from branch to branch with ease.

▽ The long legs enable the **footle** to leap nimbly through the treetops. The three fingers and four toes, all long and thin, can perch on, and grasp, the thinnest of twigs while the long jaws can root around under the leaves to catch insects hidden there. Many species of footle live in the forests of the Nearctic realm.

THE NEOTROPICAL REALM

The continent containing the Neotropical realm is almost an island, connected to the Nearctic continent by only the narrowest necks of land in the north-west. For 8,000 kilometres (5,000 m) the continent stretches north-south, from the tropics of the northern hemisphere almost to the Antarctic circle. At its widest it reaches 5,500 kilometres (3,500 m) from coast to coast just south of the equator. The realm also includes the northern land bridge last established about five million years ago, and the southernmost tail of the Nearctic continent proper. It is the barren desert conditions found here that define the barrier between the two realms. In the very south a compact string of islands connects the continent to the vast frozen continent that lies over the south pole. This continent has so little life on it that it does not constitute a zoogeographical realm.

The Neotropical realm was once the western part of the supercontinent Gondwana. This is obvious from the shape of the eastern coastline, which follows the shape of the western coastline of the Ethiopian continent exactly, showing where the two became separated in the middle Cretaceous period. Its separation from the Nearctic continent in the north was not so complete nor so final. Both the Neotropical and Nearctic continents have been moving continuously towards the west, overriding the oceanic plates there. The relative plate movements along the western coast of the two continents has meant that there has been a continual steady build-up of fold mountain chains and volcanoes along the western coasts. This build-up has continued in the ill-defined region between the continents as well, and over the millennia island chains and land bridges have been created across the sea channel, and then destroyed again. At the moment, the connection is present in the form of the land bridge in the west and a loop of volcanic islands in the east.

As with the other continents, the physical geography reflects the geological history. There are old mountains in the east, representing one edge of the old rift-valley along which the Ethiopian continent tore away during the Cretaceous per- iod. Along the entire western coast there is a new mountain range, constantly being enlarged by the reaction between the continent and the plates of the nearby ocean. Volcanoes bub- ble and stream along the length of this mountain range. The isthmus that connects this continent to the Nearctic continent can be thought of as the northward continuation of this range. Two main river basins are found in the lowlands. The central one measuring 5,827,500 square kilometres (2,250,000 sq m), represents the greatest river system in the world.

Most of the northern reaches of the realm, being within the tropics, are clothed in tropical rainforest. The huge river basin found in the broadest section of the continent is fed by the daily rain that falls here. As in most tropical forests, the favourable growing conditions mean that thousands of species of trees grow here, supporting many different kinds of tree-living animal, from the insects to the tree-living dinosaurs that feed on them. The forest floor has its own fauna of insects and insectivorous dinosaurs.

To the south, the largest areas of the lowlands are occupied by grasslands, that range from tropical in the north through temperate and then cold in the south. On these pampas live grass-eating animals; huge long-necked grazing dinosaurs like those that have existed on the continent throughout its life as an island realm. These dinosaurs are becoming rare and even extinct now, and are being replaced by other creatures that are evolving from animals that have migrated south from the Nearctic realm.

As the western mountains provide the land connection between the Neotropical and the Nearctic realms, it is hardly surprising that many of the immigrants to the realm have evolved from Nearctic mountain animals. Many of those coming south have developed into forest-living and pampas-living types. Some, however, have remained mountain creatures. Some of the largest pterosaurs in the world soar above these wild mountain chains.

There are deserts here too. The grasslands in the very south are extremely dry. They are in the rain-shadow zone of the mountains, where the prevailing winds drop all their rain on the other side of the ranges. Coastal deserts are found in a narrow strip between the mountains and the sea. Here, the cold ocean currents cause the air to cool and descend so that no rain falls. Further south on this coastal strip, however, there are moister, more fertile conditions where the wet winds blowing from the ocean have more influence on the climate than the cooling by the ocean currents.

Equator
Barren land
Tropical rainforest
Grassland and forest savanna
Grasslands
Deciduous and mixed woodland
Seasonal 'mediterranean' conditions
Desert and desert scrub
Coniferous forest

Tropical rainforest

PANGALOON

Filarmura tuburostra

An interesting example of the re-evolution of features once lost, arises in the case of the pangaloon. Its body is covered in scales, but these are not the conventional reptilian scales that were possessed by its archosaurian ancestors 200 million years ago. These scales are actually plates of fused hairs. The original scales were retained as the archosaur evolved into the primitive coelurosaur dinosaur. Then, as the coelurosaur developed and evolved into the warm-blooded arbrosaur, the scales evolved into insulating fur.

The pangaloon, which is evolved from an arbrosaur that has reverted to a ground-dwelling way of life, has redeveloped a scaly armour from masses of fur fused together. Its ground-dwelling habit evolved as the ants flourished and became widespread in the early part of the Tertiary period. As the ants evolved then so did the ant-eating adaptations of animals like the pangaloon. The typical, long toothy jaws of the arbrosaur have been replaced, in the pangaloon, by a tubular snout, along which lies a long sticky tongue that can be thrust out to a distance greater than the length of the head. It can push its snout down ant burrows and lap up the ants that adhere to the tongue's stickiness. The middle claw of the forelimb has developed into a strong hook that can tear into ants' nests to expose their tunnels and chambers. The nostrils are high up on the face and can be closed for protection.

Tropical rainforest – swamps

WATERGULP

Fluvisaurus hauristus

The waters of the vast network of channels and tributaries that pour into the great river flowing eastwards along the equator are alive with aquatic animals of one kind or another. The forest reaches over the quiet bayous and backwaters, steeping them in a murky green gloom. Water plants root in the deep mud, reach upwards through the turbid waters and spread their leaves out over the placid surface. The undersides of the floating leaves support colonies of water snails and other aquatic invertebrates, and these are eaten both by fish, and by lightly built animals that crawl over the tops of the leaves and probe downwards. The floating leaves themselves are eaten by huge, placid, slow-moving water beasts that swim lazily in the shallows and engulf huge mouthfuls of the vegetation.

The largest of these river browsers is the watergulp, evolved from the swift-footed hypsilophodonts that were so successful in Cretaceous times. The powerful hind legs have since evolved into paddles, as have the versatile five-fingered forelimbs. These are used mainly for stabilizing the animal's great bulk, and the two claws on the forepaddles are useful for grubbing around the muddy roots. Most of the swimming action is achieved by the tail. The stiff rod-like tail of the ancestral hypsilophodont has become more flexible, and a tall leathery fin along the upper surface can produce powerful swimming strokes.

a

b

The **pangaloon** has a long sticky tongue (**a**) which it extends into ant burrows. The dense scaly armour of the pangaloon is not a protection against ants but guards against attacks by predatory reptiles that live on the tropical forest floor. The pangaloon is a slow-moving animal and unlike its relatives, it cannot take refuge in trees. When it is threatened, the pangaloon can curl the paddle-shaped tail (**b**) beneath its body to protect its soft underparts from the stings and biting jaws of the large ants.

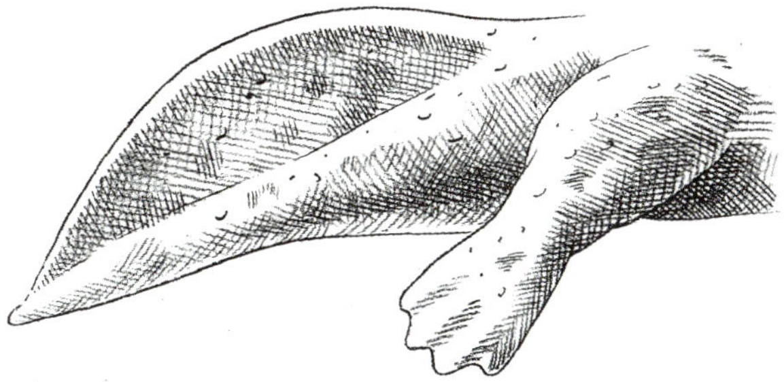

The evolution of the legs and tail of the **watergulp** mean that it can move only in water. Its eyes and nostrils are high on the head, allowing it to see above the water surface when submerged. The mouth is broad, and the sharp-edged horny beak can cut through the stems and leaves of the toughest water plants. At about 2.5 metres (8 ft) long it is one of the largest river animals. Each watergulp also supports a large colony of parasites and other companions that feed on algae growing on its flanks, and on small creatures disturbed from the mud by its passage.

△ As in many aquatic vertebrates, the **watergulp's** ribs are very heavy, giving it enough weight to keep it submerged. It also swallows stones from time to time to adjust its buoyancy.

▷ The nectar-eating **gimp** feeds rather like moths and bees, pushing its snout down into the base of long flowers to reach the nectar. As a result, the snout emerges covered with pollen which the gimp then unwittingly transfers to another flower, thus fertilizing it. The animal's shape is similar to that of the small arbrosaurs but it is not as fast-moving. Its forelimbs can grasp flower stems. Its body is very small, since nectar needs very little digestion and requires only a small gut.

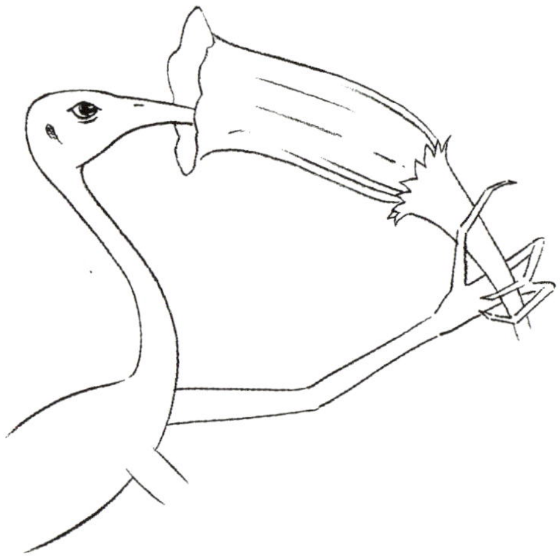

Tropical rainforest

GIMP

Melexsorbius parvus

As in tropical forests all round the equator, the trees of the Neotropical realm are full of numerous kinds of animals. They support many more different types of animal than the trees of the forests in the temperate or cold zones. With the constant high temperatures and the daily rainfall, vegetation flourishes and several hundred different species of tree thrive in a very small area. A thick canopy consisting of intermeshed boughs of the tallest trees spreads out to catch the sunlight. Epiphytes and dangling creepers festoon these high branches. As a result, there are many different kinds of leaves, flowers and fruits available above the gloom and roots of the forest floor. Many varieties of animals have evolved to exploit these potential foodstuffs. Generally, the tree-living animals of the Neotropical realm are smaller than those of the tropical forests in the rest of the world. The airy canopy is alive with little arbrosaurs adapted to feed on the thousands upon thousands of different types of insects that make their homes there.

Some of the arbrosaurs, however, have abandoned their insectivorous way of life and have evolved into totally new forms. The nectar-sucking gimp, for instance, is a tiny animal, no more than 20 centimetres (8 in) long including the slender delicate head, and eats nothing but nectar. Its snout has evolved into a long tube which acts as a rigid sheath for an extendable nectar-gathering tongue. These features are similar to those of the ant-eating adaptation of the pangaloon (page 68), and it is likely that both animals evolved from the same group of arbrosaurs that crossed from the Nearctic realm early in the Tertiary period. The tubular snout may be seen as an example of pre-adaptation where a feature evolves spontaneously and is then retained because it is perfectly suited for a particular purpose.

▷ The **scaly glider** has a set of paddle-like scales, growing horizontally from each side of the body and spreading out to form gliding wings. Muscles in the flanks and attached to the ribs give the scales some limited mobility, allowing the bright undersides to be exposed or hidden at will.

△ Many species of **gimp** exist in the Neotropical forest. Each one eats the nectar of a particular species of flower and has a correspondingly different shaped snout. A more obvious difference between species is the pattern on the back. This can range from spots, to stripes, or large patches of colour.

Tropical rainforest

SCALY GLIDER

Pennasaurus volans

Bright flashes of colour appear in the heights, up among the branches, creepers and trunks of the Neotropical canopy. To and fro, the colours dart, like large butterflies, swooping and wheeling in the breezes, then flopping onto a branch and blending instantly with the mottled bark in the dappled sunshine. These are not butterflies, however, but some of the smallest dinosaurs that have ever existed. Descended from the same coelurosaur stock as the arbrosaurs, the scaly gliders adapted to a gliding life with the evolution of the flowering plants and hence, the butterflies. The latter represent the gliders' principal food. A controlled dive and glide through the air can bring a scaly glider close to a butterfly at rest on a flower, and, with a quick snatch of pointed jaws, the insect is caught and the glider sails on to rest and digest.

Flying and gliding animals frequently evolve in tropical forests. With the network of massive boughs and close-growing branches, animals can easily jump from one tree to another. The more aerodynamic the animal's shape, the further it can jump. The conditions are perfect for the evolution of gliding structures that can carry an animal over even longer spaces. In the case of the scaly glider, these structures consist of broad scales projecting sideways from the body. In flight, they turn the animal into an aerodynamic skimmer, while at rest, they help to camouflage it against the tree-bark. They are brightly hued beneath, but drab when seen from above.

▷ With a total length of only 30 centimetres (12 in), including the whip-like tail, and a 'wing' span of 25 centimetres (10 in), the **scaly glider** is the smallest and most lightweight of all the dinosaurs.

Grassland – pampas

TURTOSAUR

Turotosaurus armatus

The Neotropical continent has had an interesting history. After it had broken away from the ancestral supercontinent it remained an island continent for most of its life. It connected only occasionally to the continent of the Nearctic realm in the north, as it is today. At the end of the Cretaceous period the duckbilled dinosaurs spread over most of the continents and established themselves as the principal large herbivores, replacing the great four-footed sauropods. The sauropods still exist, but mostly in places where the duckbills never gained a foothold, such as the Neotropical continent. Isolated from the influences of migration from other continents, the sauropods evolved in their own way.

The titanosaur family – the only surviving family of sauropod – had begun to develop armour as early as the late Cretaceous period. They continued to do so in the Neotropical continent and, in the late Cenozoic, with the arrival of the new meat-eaters, the armour developed into very flamboyant and spectacular forms. The turtosaur has a solid shield of armour over its back and flanks. The tail and neck are also shielded, the tail by overlapping rings of horn, and the neck by jointed horn sections. The head is covered with a continuous horny shield. The jaws have no teeth, but the mouth edges of the horn shield are sharp and are ideal for cropping grass.

Sauropod Evolution

Throughout their evolution the sauropods seemed to diminish in size. They reached their largest size in the late Jurassic, and then, towards the late Cretaceous, a number of small forms appeared. This trend continued until there were a number of quite lightweight sauropods in the Neotropical continent in Cenozoic times. Some were quite fleet of foot and would run gracefully in herds across the pampas.

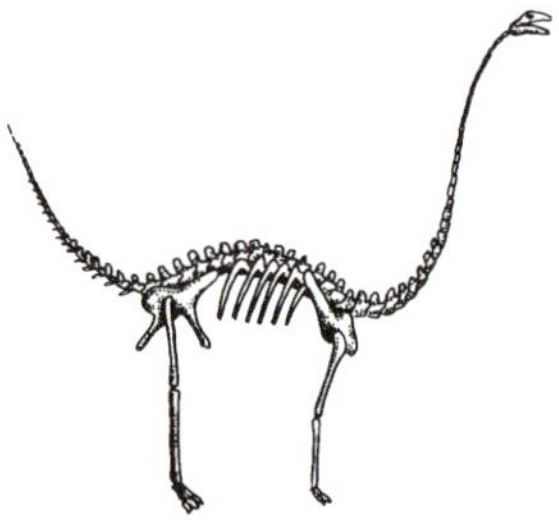

▷ The armour of the **turtosaur** is formed from bone masses that grow in the skin and are covered by a layer of horn. The back armour is a solid mosaic of small skin bones, with much larger bones spread between them. These larger bones provide the bases for a series of conical spikes. Huge hip bones and shoulder bones support the weight. There is no armour on the legs or the underside of the tail. When danger threatens the animal flops down on its belly and presents an impenetrable shield of bone and horn to its enemy.

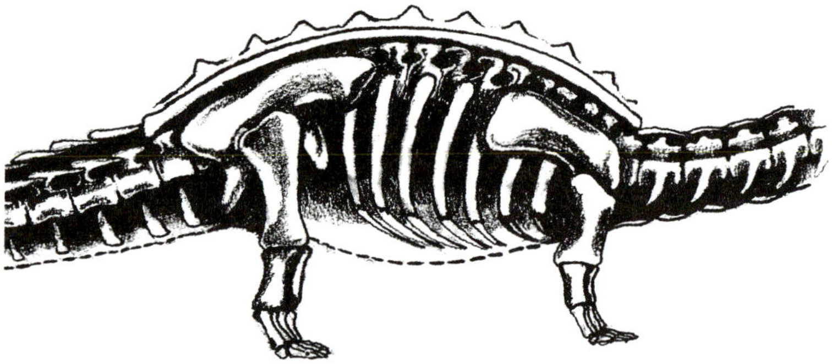

The long neck and the trunk of the **lumber** enable it to graze over a large area without moving far. The neck is quite stiff but can be swung from side to side at the shoulders. Towards the head it becomes more flexible. The teeth are confined to the front of its mouth (**a**) and are adapted for cropping grass. The muscles of the short trunk are anchored on broad plates of bone at each side of the enormous nostril openings in the skull (**b**). The trunk is a feeding organ used to grasp and uproot large bunches of grass. It is also a breathing organ, but the position of the external nostrils do not interfere with its feeding action (**c**).

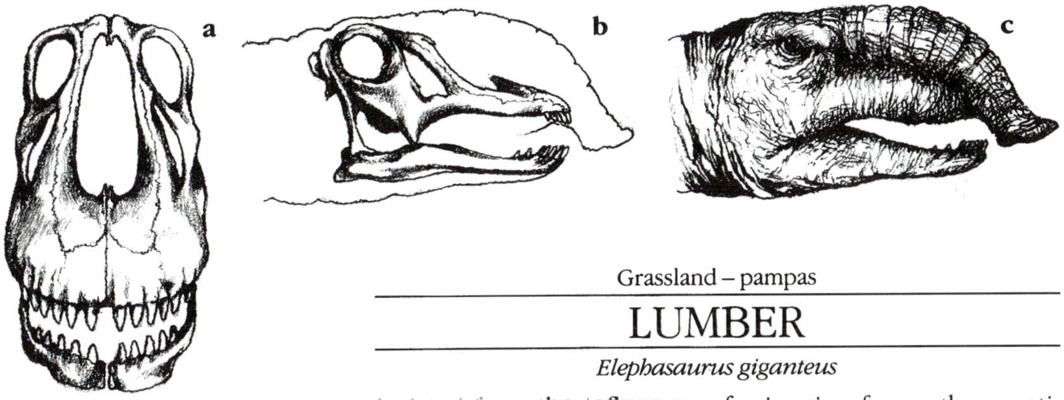

Grassland – pampas

LUMBER

Elephasaurus giganteus

Isolated from the influences of migration from other continents, the sauropods of the Neotropical continent evolved in their own way to cope with such changing conditions as the spread of the grasslands. The lumber is a grassland-living sauropod that has evolved from a tree browser into a grass-grazer. During the Tertiary period many strange grassland-dwelling sauropods developed to live in the newly evolved pampas, including a number of long-legged, running forms. These were very vulnerable to the swift-footed carnivorous dinosaurs (pages 74–5) that spread across the continent about four million years ago, when the present land bridge was established to the Nearctic continent to the north. The long-legged sauropods were fast runners, but they were nowhere near as fast as the meat-eaters and were eventually made extinct.

The sauropods that survived did so because they evolved defensive strategies. The turtosaur, for instance, evolved armour. Others, such as the lumber, took refuge in mere size. The lumber is the largest land animal alive today, with a length of 25 metres (80 ft) and a weight of 70 tonnes. It cannot raise its neck far above its shoulders, but it can sweep round and reach vast areas of grass without moving its feet. Its teeth are confined to the front of its mouth and are hard-wearing, adapted for cropping grass. The trunk helps to pull bunches of grass towards the mouth where they are torn off and swallowed. The skin is thick and leathery, quite resistant to meat-eaters' claws and teeth, and most carnivores leave the huge creature well alone.

Grassland – pampas

CUTLASSTOOTH

Caedosaurus gladiadens

Many animals flooded into the Neotropical continent about four million years ago, when the string of volcanic islands linking with the Nearctic continent to the north, fused to form a land bridge. The animals quickly evolved and developed to fill the ecological niches they found. In most instances the animals that came south were more successful and began to replace the indigenous animals that had dominated the continent for the previous 50 million years. The reason for this may be that the indigenous animals had evolved few variations in that period of time. The continent itself had not shifted substantially and there had been little change in the environment and the climate except for the spread of the grasslands. There was no necessity then, for the animals to evolve, and they had settled into a long period of decadence.

The animals from the north, on the other hand, had suffered millions of years of change, with changes in climate and interchanges of animals with the Palaearctic realm, and had been evolving vigorously. When they reached the Neotropical continent they were more able to adapt to the conditions found there and, as a result, the native animals suffered. The most noticeable instance is found among the meat-eaters. The first of the northern animals to reach the continent were the mountain-leapers (page 61). Being skilled at traversing high mountain ranges, they were well able to cross the mountainous neck of land that connected the two continents and spread into many of the ecological niches. Then there migrated a group of ornithomimids, which quickly evolved a distinctive lifestyle. They grew big and heavy, and retained their juvenile teeth through life. But these teeth developed into huge slashing fangs, enabling these animals to become the main hunters of the time and the area.

One such is the cutlasstooth which has evolved huge cutting teeth that enable it to prey on the large sauropods like the lumber (page 73). Hunting in packs, cutlasstooths can set upon a lumber and slash it until it bleeds to death.

b

a

△ The **cutlasstooth's** dental arrangement is unique among dinosaurs. The first tooth of each upper row is a long, curved slashing weapon (**a**). Those behind it are continually growing. Once the front tooth is broken and discarded there is another to take its place (**b**).

Grassland – pampas

GOURMAND

Ganeosaurus tardus

As a rule the bigger the animal, the shorter is its species life-span. This holds particularly true for the biggest of the meat-eating dinosaurs.

During the Jurassic period the biggest of the meat-eaters were the allosaurs, members of a larger group sometimes called the carnosaurs. An offshoot from these animals was the carcharodontosaur group, which flourished well into the Cretaceous and still exist today but only as specialized types in the islands of the Ethiopian realm (pp.38–41).

Another group of huge meat-eaters from the late Jurassic and early Cretaceous were the spinosaurs, with long narrow heads and spectacular sails on their backs, but these were all extinct by the end of the Cretaceous period and have no descendants today.

Perhaps the most spectacular of the big theropods were the tyrannosaurs. These evolved quite late in the dinosaur history and existed for a while as medium-sized beasts, in the shadow of these other big animals. But towards the end of the Cretaceous they evolved into massive hunters like *Tarbosaurus*, *Daspletosaurus* and *Tyrannosaurus.*

While the huge tyrannosaurs dominated the northern continents, another group expanded in the south. Evolved from the primitive ceratosaur group of theropods, the abelisaurs soon came to dominate these southern landmasses. As the tyrannosaurs declined, the abelisaurs spread northwards to take their places, and are now the dominant big meat-eaters throughout the world.

The biggest of these is the gourmand. It lives on the pampas of the neotropical realm, the area where its ancestors first evolved. The big tyrannosaurs were remarkable for their short arms, but those of the biggest abelisaurs were tiny. The gourmand has taken this to an extreme and the arms have disappeared completely. The animal is a scavenger and moves slowly across the open plains swallowing whole any carcass it comes across. The skull is very flexible and the jaws can unhinge, like those of a snake, to swallow its prey. After feeding it digests its meal, lying motionless in the grass and relying on the armour of its back for protection.

d

c

△ The jaws of the **cutlasstooth** can open remarkably wide, so that the lower jaw is well away from the cutting teeth when the heavy head is thrust downwards in the slashing stroke (**c**). The undersides of the large teeth are serrated and work against the lower teeth as carnassial (meat-shearing) blades for ripping and chewing flesh when the animal is eating. When hunting, a pack of four or five can inflict deep bleeding wounds on the flanks of their prey (**d**), and then wait until it bleeds to death.

◁ The slow-moving **gourmand** is protected from other meat-eaters by a back armour of bony plates sheathed in horn. Its scavenging life style is very different from that of its ancestors.

▷ The **dip** is a patient animal and sits by the pools of mountain waterfalls waiting for fish. It is able to judge its distance from its target despite the distorting effect of the water, and darts its long head downwards, cleaving the water and pinning the slippery fish between its many sharp teeth.

Barren land – mountains

DIP

Harundosaurus montanus

The western mountain range that forms the backbone of the Neotropical continent traps moist air blowing in from the ocean and is always drenched in rain. The precipitation produces streams that cascade steeply back to the ocean on the western side, or form small rivers that develop into the tributaries of the huge rivers that meander across the lowlands to the east. These streams are full of different types of fish and invertebrates. The fish are eaten by various animals including birds, pterosaurs and specialized hunting dinosaurs.

One of the most interesting animals is the dip, a fish-eating dinosaur evolved from Nearactic invaders. Like many of the hunting dinosaurs of the Neotropical continent, the dip is descended from the ornithomimids (p.61) that migrated southwards when the land bridge to the Nearctic continent was established about four million years ago. It retains the long silky fur of its ancestors that is necessary to protect it from the harsh mountain climates. The head has evolved the shape that has developed in many groups of fish-eating animals. The jaws are long and narrow and furnished with many fine-pointed teeth. The eyes are set so that they are directed forward and give stereoscopic vision. Like the mountain leaper, the dip is very fleet of foot and can run acrobatically along the sides of crags to escape the predatory birds and pterosaurs that abound in the region.

▽ The **harridan** is a mountain pterosaur, evolved to soar in the mountain air currents. It is equally at home on the ground, however. The wings and leg membranes can fold away when it is at rest, and the harridan can walk about the crags and ledges like any other biped.

▷ With its keen stereoscopic vision, the **harridan** can accurately swoop down and snatch unsuspecting small animals living on the ledges far below the mountain peaks.

△ The **harridan** has a wingspan of over 5 metres (17 ft). In addition to the wing membranes it possesses membranes on its hind legs. Each of these is supported by a very long first toe, and is used to control the animal in flight.

Barren land – mountains

HARRIDAN

Harpyia latala

Between the mountain peaks and above the valleys of the western mountain range soar a number of large birds and pterosaurs. They circle in the winds rising from the valley heads and the updraughts from warm hill slopes. One of the largest of these soaring creatures is the harridan, a pterosaur. It can manoeuvre expertly while in flight. The air sweeping over its broad wings is controlled by additional membranes attached to the second and third fingers of the hand. These two fingers are elongated but not as much as the fourth finger that in all pterosaurs, supports the main flying membrane. The first finger carries a claw which is used when the animal is crawling on cliff faces. Its head is almost mammal-like, with an elongated snout and sharp teeth at the front. The eyes are directed forward and give a good stereoscopic vision.

This pterosaur is a solitary animal. It mates for life and nests on high pinnacles in the mountains, rearing no more than two young each year. With its warm-blooded physiology, and its carnivorous hunting habits, it is very similar to the predatory birds that are found in the mountainous areas of other realms.

The harridan has very keen eyesight and can spot small animals scampering about on the ledges hundreds of metres below. It then swoops out of the sky and pounces accurately on its food. Its prey usually consists of the small mammals that have adopted a mountain habitat.

THE ORIENTAL REALM

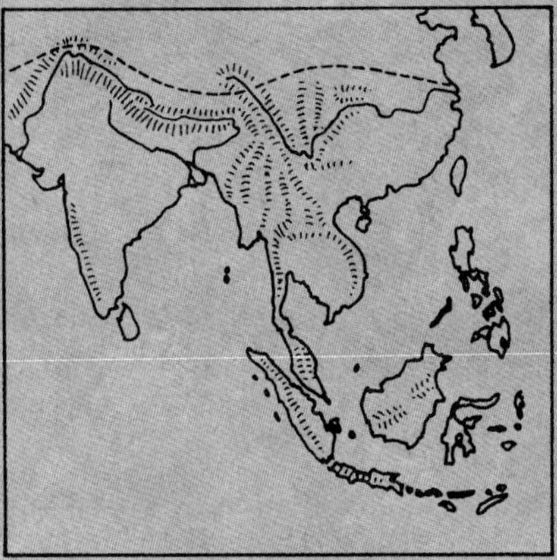

The Oriental realm is the smallest of the zoogeographic realms, and, in many respects, it is the most difficult to define. Geographically it is part of the great northern continent and is that section of it that lies to the south of the huge barrier of mountain chains that crease up its south-eastern corner. However, it also encompasses the cluster of jungle peninsulas and volcanic island archipelagoes that reach out across the ocean towards the Australasian realm in the southeast. The most extensive flat area of the realm is a triangular peninsula 2,000 kilometres (1,250 m) wide by 2,500 kilometres (1,550 m) long, consisting of an old fragment of Gondwana. Most of the rest of the realm is quite new and is formed of fold mountains. The northern boundary is well defined by the ramparts of the mountains. To the west, the boundary lies in a desert region that is continuous with that in the north of the Ethiopian realm. The south-eastern boundary is not well defined. It lies somewhere in the sounds and straits between the strings of islands. Many animals can move from one island to another by swimming across the intervening sea areas, or by drifting on mats of vegetation, or even by walking across when the sea level permits. Some Australasian animals have penetrated well into the Oriental realm in this way, and many Oriental animals have similarly spread into Australasian territory. It is difficult to place a realistic boundary between the two realms here.

The fragment of Gondwana that formed the main peninsula did not break away from its southern supercontinent until the late Jurassic. Since then it has moved northwards across the ancient Tethys Sea, out of the southern hemisphere and into the northern, and eventually collided with the northern continent as recently as 30 million years ago. The rapid approach of one continent to the other crumpled up the seafloor sediments between them, and the final collision crushed these up into the greatest mountain range on the face of the Earth today. The disturbance caused by this collision was wide-spread. Its ripples produced the mountain ranges to the east, those that reach out their crests into the sea to form the peninsulas. The oceanic plate that carried the piece of Gondwana northwards is still forcing its way beneath its neighbour, and producing the chains of volcanic islands that lie between the northern continent and that of Australasia. In geological terms this area is constantly changing.

The Oriental realm is one of mountains and islands, and its habitats reflect this. It is something of a zoological crossroads, many of its animals having arrived from neighbouring realms. Deciduous forest and wooded grasslands lie on the great triangular peninsula. These are grazed and browsed by the descendants of the old Gondwana fauna - big long-necked plant-eaters that are only found on the other fragments of the Gondwana supercontinent. These share the plains with creatures more typical of the Palaearctic and Ethiopian realms that have moved in and adapted well to the environment.

The rivers that pour off the mountain ranges produce lush, forested deltas where they flow into the broad bays of the ocean. The deeply indented coastline means that there are few permanent ocean currents and the river sediments can build up into huge deltas here. Trees, such as mangroves, that can withstand flooding by both fresh and salt water, establish themselves on mudbanks and trap more silt by their root systems, so building up more areas of land along the coasts. The forests receive heavy rainfall during the summer months because of the monsoon climate. The warming continent to the north produces rising air masses that draw in moist winds from the sea. In the winter the opposite is the case and dry winds blow outwards from the arid interior. The forests and deltas are home to thousands of different tree-living and aquatic dinosaurs, many of them evolved from creatures that originated in the Ethiopian, Palaearctic and Australasian realms.

The mountains to the north are the highest on Earth. They are also some of the steepest. High mountain peaks and glaciated ridges can drop to tropical forested ravines and valleys within a few kilometres, passing through narrow zones of alpine plants, and rhododendron and bamboo forests on the way. The alpine plants found at high altitudes are browsed by nimble-footed dinosaurs, only a few kilometres from where tropical forest dinosaurs browse the steamy forests in the deep valleys.

Deserts lie in the west, defining the boundary to the realm. Across these have travelled a number of dinosaurs adapted to arid conditions, which have then gone on to develop quite easily into forest-dwellers.

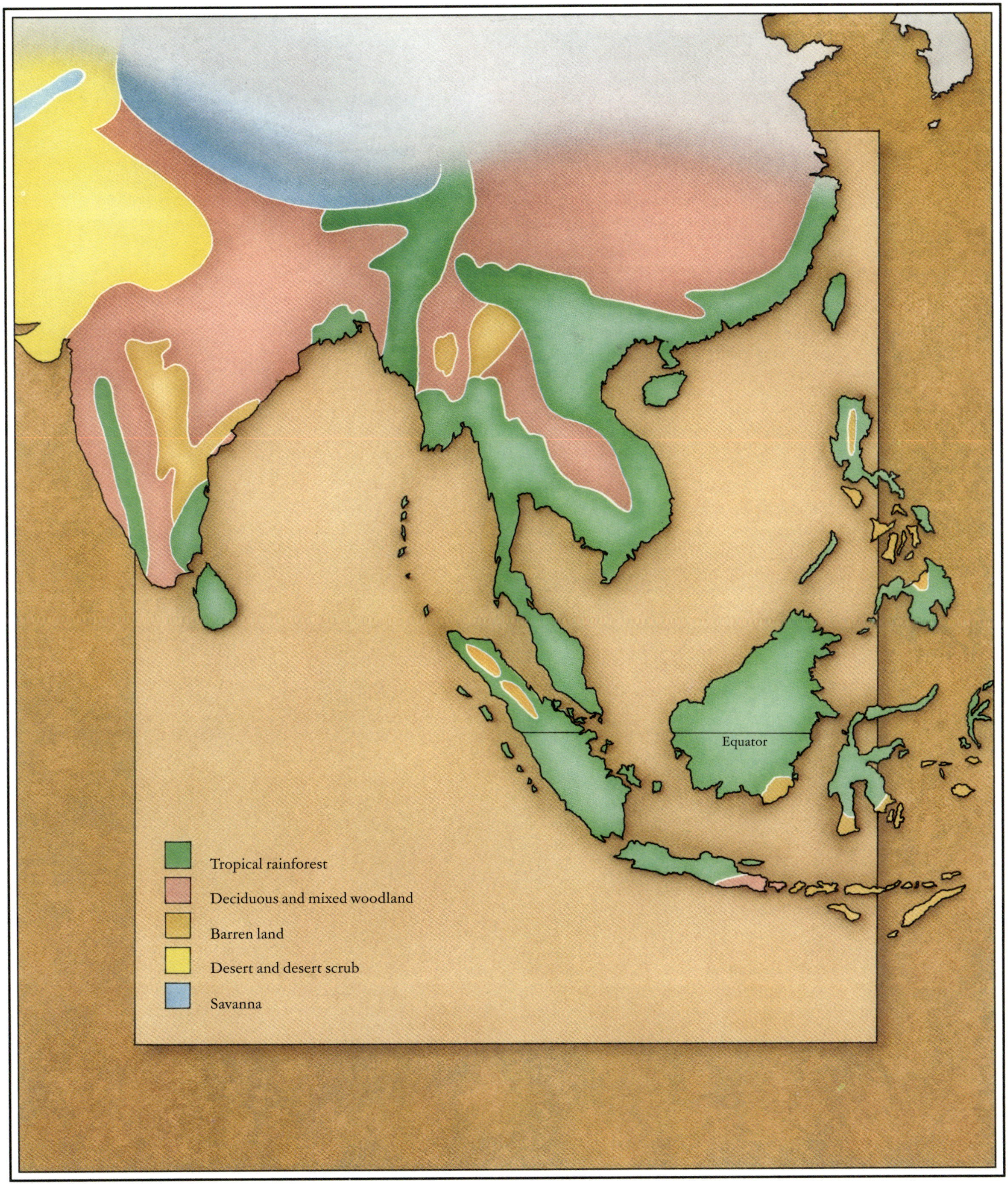
Equator
Tropical rainforest
Deciduous and mixed woodland
Barren land
Desert and desert scrub
Savanna

Savanna-plains

RAJAPHANT

Gregisaurus titanops

During the Cenozoic a large triangular piece of ancient Gondwana, over 3,000 kilometres (1,900 m) long and 2,500 kilometres (1,550 m) broad, moved north-eastwards across the Tethys ocean and fused with the northern continent. It threw up the greatest mountain range on Earth along the join. The Gondwana animal life continued on this new triangular peninsula with little change, isolated from the northern continent by the mountain range.

The largest animal now on the sub-continent is the rajaphant, evolved from the Cretaceous titanosaurs. It wanders in small herds, grazing the yellow grasses of the central plains. Although its physical appearance has not changed very much from that of its ancestors, it has developed a complex social structure for life in the exposed grasslands. It has also evolved a feeding mechanism that allows it to eat tough grasses as well as the soft leaves of the bushes and trees. The mouth is broad and the teeth are sharp enough to crop grass, but not strong enough to chew it. The grass is swallowed in vast quantities and collects in an enormous muscular gizzard, a metre (3 ft) wide, in the forward part of the stomach. There it is pounded and crushed to a digestible pulp by masses of hard stones that are deliberately swallowed by the rajaphant from time to time.

▷ The **rajaphant** herds have evolved defensive habits and take great care to protect their vulnerable young against the predators. The adults cluster round and lash out at the swooping hunters with their long necks and teeth, and their heavy whip-like tails.

◁ The open plains and grasslands are visited by many species of predatory birds and hunting pterosaurs, that soar and circle in the hot skies searching for likely victims. Anything moving on the exposed ground is a potential meal.

△ Each **rajaphant** herd has a strict social structure, and this is especially evident while the herd is on the march across the grasslands in the rainy season. Out in front is an old bull, the undisputed leader of the herd. The younger bulls march along the flanks in a defensive formation. The females shepherd the calves in a knot in the centre of the herd.

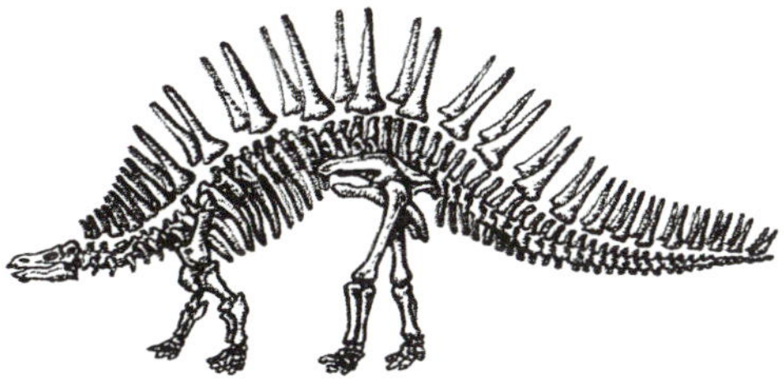

Of all the thriving dinosaur groups of Mesozoic times there is one missing from the present-day fauna – the stegosaurs. These plated dinosaurs had varying arrangements of vertical plates or spines on the back. They were the main members of the thyreophora in the Jurassic but disappeared at the beginning of the Cretaceous, their places taken by the armoured ankylosaurs.

◁ The **hanuhan** lives in small family groups, usually scattered over a wide area of mountainside. The food is so sparse at altitudes of about 4,000 metres (13,500 ft) that larger concentrations of animals would not survive. The agility of the hanuhans enables them to move from one bleak pasture to another with ease, hopping and leaping from rock to rock, balanced by the stiff rod of the tail.

Barren land – mountains

HANUHAN

Grimposaurus pernipes

Up towards the snowline of the greatest mountain range on Earth, the range that divides the Oriental from the Palaearctic realm, there is very little to support life. Yet even these bleak regions have a number of highly adapted animals.

The hanuhan is a dinosaur that has evolved to live in these harsh conditions. Its adaptations are similar to those of the balaclav (page 61): deep layers of fat for insulation, and strong claws and beak for scraping the sparse plant material – mosses, lichens and alpine plants – from the rocks crannies. Like the balaclav, too, it has evolved from the successful hypsilophodonts of the Cretaceous period. As such, it has probably come into the Oriental realm across the vast mountain range from the Palaearctic realm, rather than having been brought up from Gondwana in the south on the drifting continent. It is a very nimble animal, sure-footed on crags and confident on the narrowest of ledges. The brain has developed well to control balance and muscular coordination. It could be that the hanuhan evolved from tree-living hypsilophodonts before developing adaptations suitable for its mountaineering life style.

△ The large size of the **taddey** – up to 2 metres (6 ft) long without the tail – and its lazy appearance, makes it quite different from its agile hypsilophodont ancestors.

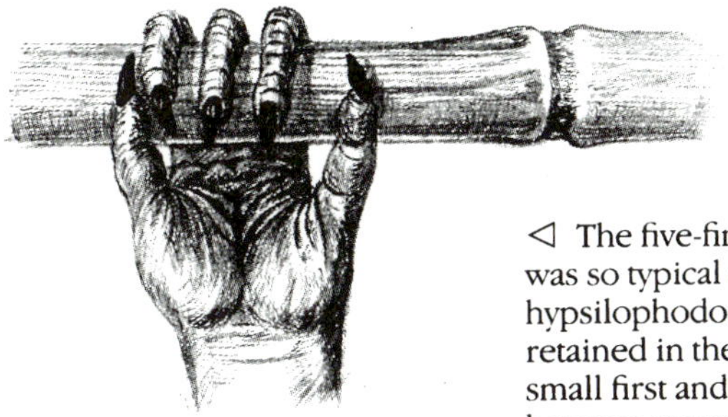

◁ The five-fingered hand that was so typical of the hypsilophodonts has been retained in the **taddey.** The small first and fifth fingers have become versatile and opposable, acting as small thumbs that can fold over the palm. The resultant hand is able to firmly grasp the bamboo stems which are taddey's main food. The beak is used to scrape leaves and shoots from the woody canes.

Tropical rainforest – mountains

NUMBSKULL

Sphaeracephalus riparus

The great mountain range is young, but it is already deeply weathered and eroded. Annually, melting snows from the higher slopes and the constantly decaying glaciers produce torrents of water that form streams cascading from the heights towards the distant plains and ocean. At the same time the mountain range is continually pushed upwards by the inexorable movements of the Earth's crust. As the mountains rise, the river gorges deepen. This has led to an unusual circumstance, in which a major eastward-flowing river on the northern flanks of the range has turned south and cut right across the mountain system to reach the sea in a mass of deltas to the south. Cutting through the range, it has formed a gorge that is the deepest on Earth. The lush vegetation that grows along the narrow banks of the river in the depths of the gorge supports its own wildlife. The most typical animal is the numbskull, one of the pachycephalosaurs, or bone-headed dinosaurs. It has changed very little since the group evolved in Cretaceous times, and it still pursues the same herbivorous life style. It lives in herds and family groups in the river gorge, with the males periodically sparring with one another for leadership of the herd. The numbskull is bipedal and roams through the tropical forest found in the gorge, maintaining balance with the strong tail.

◁ The late Cretaceous period pachycephalosaurs spread worldwide. Many developed into new forms to exist in new environments (pages 44–5) but some, like the **numbskull,** remained quite conservative.

Barren land – mountains

TADDEY

Multipollex moffati

Well below the snowline, beyond the bare rocks and sparse alpine plants that are the habitat of the hanuhan, the mountains fall away, slope after slope, into the lower flanks and the foothills. On the slopes that reach downwards from about 4,000 metres (13,500 ft) to the misty regions of about 2,000 metres (7,000 ft), the climate is equable and the vegetation more prolific. In the higher reaches the alpine meadows give way to rhododendron scrub, and then to belts of bamboo forest. The greater variety and volume of plants means that more animals live here than further up. One of the largest of these is the taddey. Another descendant of the hypsilophodonts, probably from the same stock as gave rise to the hanuhan, the taddey is a slow-moving forest animal that feeds almost entirely on bamboo shoots. Several species and subspecies of taddey exist in the bamboo thickets, each one being isolated from the others on particular ranges of foothills. It is a large and heavy animal, unlike its hypsilophodont ancestors, and is able to maintain its slow way of life in the absence of ground-living predators in the rhododendron and bamboo forests.

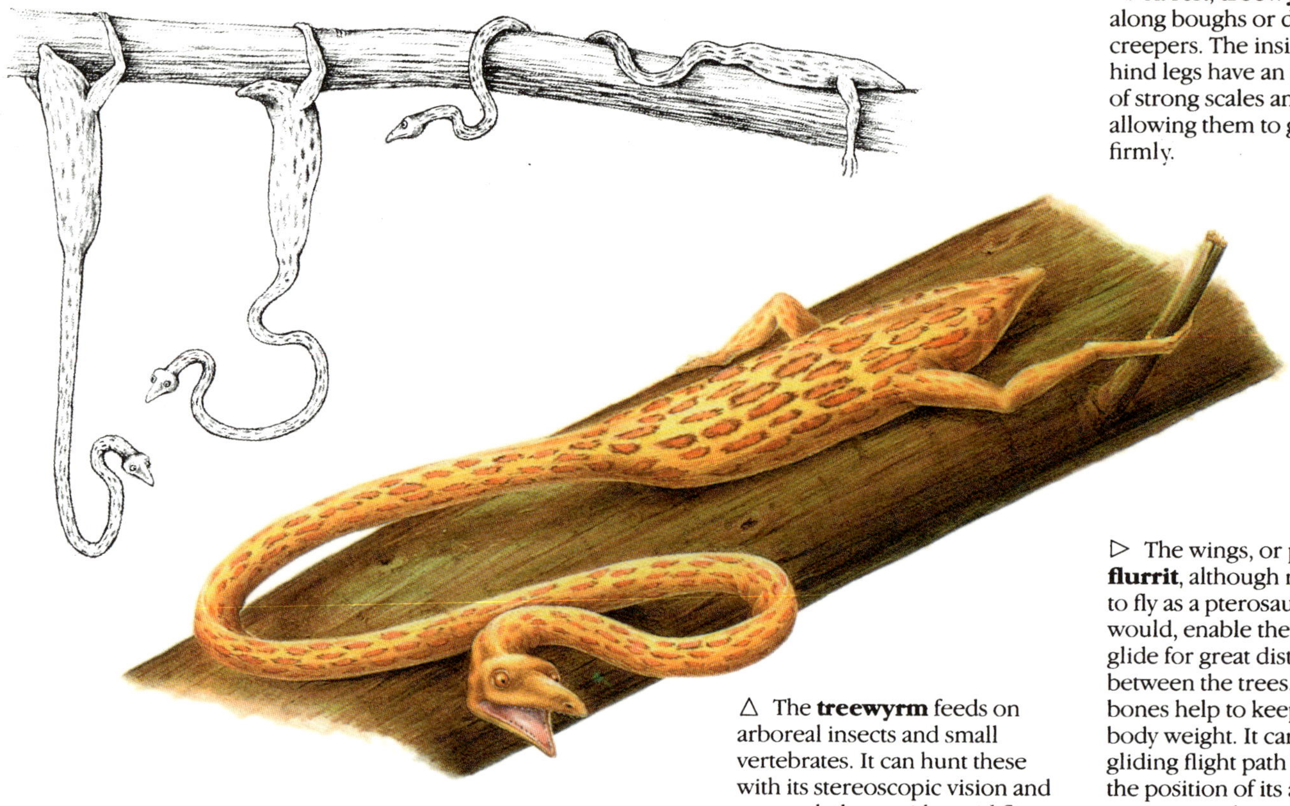

◁ At rest, **treewyrms** lie along boughs or dangle like creepers. The insides of their hind legs have an arrangement of strong scales angled inwards allowing them to grip branches firmly.

△ The **treewyrm** feeds on arboreal insects and small vertebrates. It can hunt these with its stereoscopic vision and can catch them with rapid flips of its long neck and snaps of its jaws.

▷ The wings, or patagia, of the **flurrit**, although not allowing it to fly as a pterosaur's wings would, enable the flurrit to glide for great distances between the trees. Its hollow bones help to keep down the body weight. It can control its gliding flight path by altering the position of its arms and by swinging its long tail. When hunting or feeding *below*, the flurrit folds its patagia away.

Tropical rainforest

TREEWYRM

Arbroserperus longus

The bulk of the lowland areas of the Oriental realm consists of tropical forest. The climate, known as a monsoon climate, supports forest habitats. During the summer the constant heat over the northern continent gives rise to low pressure areas that bring moist winds in from the ocean that lies to the south. A rainy season ensues. During the winter the continent cools and dry air blows out towards the ocean. The mainland forests thrive in this seasonal rainfall, but the forests of the island chains to the south-east receive the wet winds all year round – both northerly and southerly winds blowing over sea areas before reaching them – and are particularly luxurious. Tree life is abundant, and many species of treewyrm exist in the tropical forest.

The treewyrm is directly descended from the desert burrowing wyrms of the Ethiopian realm (page 36). Although the sinuous shape developed in response to a burrowing life style, it became suitable for many other ways of life. The extra ribs along the neck can be moved in waves, like the legs of a centipede, and allow the animal to crawl up very steep slopes. The bulbous body is pushed along behind by the hind legs. The treewyrm feeds on small insects and vertebrates.

The distribution of treewyrms is widespread, especially in the tropical forests of the Oriental realm, the animals having moved there across the deserts that form the boundary to the realm to the south-west of the great mountain chain.

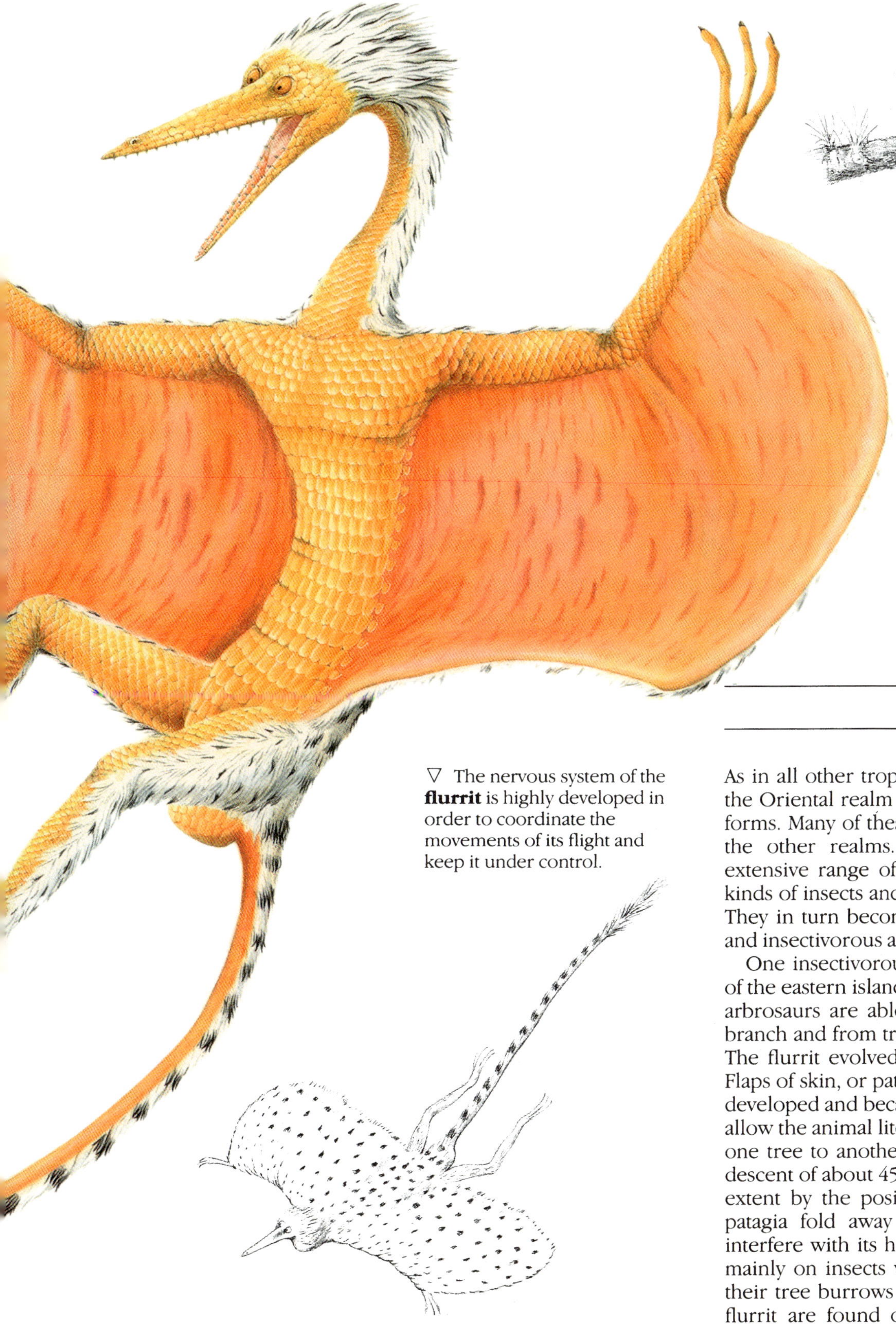

△ The **flurrit** is very small, its body and head being about 30 centimetres (1 ft) long. The underside of the wings has a very bright pattern, invisible when the wings are folded, and this is used for display and intimidation.

▽ The nervous system of the **flurrit** is highly developed in order to coordinate the movements of its flight and keep it under control.

Tropical rainforest

FLURRIT

Labisaurus alatus

As in all other tropical forest areas the tree-living animals of the Oriental realm have evolved into a vast variety of strange forms. Many of these have developed from the arbrosaurs of the other realms. The tropical forests here support an extensive range of plants, and so there are many different kinds of insects and plant-living creatures found in the forest. They in turn become prey for a wide variety of carnivorous and insectivorous animals.

One insectivorous animal that is found only in the forests of the eastern islands is the gliding arbrosaur, the flurrit. Most arbrosaurs are able to leap great distances from branch to branch and from tree to tree, in order to look for more prey. The flurrit evolved its gliding habit from these beginnings. Flaps of skin, or patagia, between the forelimbs and the body, developed and became aerodynamic structures. These do not allow the animal literally to fly but they enable it to glide from one tree to another. The flurrit's glide path has an angle of descent of about 45 degrees which can be controlled to some extent by the positions of the arms. When not in use the patagia fold away against the animal's sides and do not interfere with its hunting. Like the other arbrosaurs, it feeds mainly on insects which it catches by winkling them out of their tree burrows with its long fingers. Different species of flurrit are found on other islands. They are distinguished from one another by their markings.

▷ The **paraso,** unlike other fishing pterosaurs, is a solitary animal, flying and hunting on its own (**a**). Its head and jaws are very slim, with many tiny sharp teeth (**b**). These teeth are typical of a fishing animal, being able to grip slippery fish firmly.

a

Mixed woodland – swamps

PARASO

Umbrala solitara

Vast areas of tropical swamp and deltas are found in the Oriental realm, where great rivers, fed by the snow and glaciers of the mighty mountains, wander across the lowlands and empty into the ocean. Mangrove swamps develop at the sea's edge on the mudbanks built up by debris deposited by the rivers. The mangrove swamp usually blends into the tropical forest on the drier land, and many of the tree creatures that live here are the same as those further inland.

The swamps are home to a great many fish and aquatic animals, particularly those that can withstand the conditions of both salt water and fresh water. They also support vast numbers of wading birds and pterosaurs that feed on the fish and the mud creatures. The paraso is one of the oddest of such animals. It is a pterosaur that hunts in the shallow lagoons and backwaters of the swamps. It is quite large, with a wingspan of about 3 metres (10 ft). The wings are brightly coloured and make a dazzling display as the pterosaur launches itself into the air and flies off over the swamp. The wings, however, have another function. They are used as a kind of a trap in order to lure and catch fish in a way that is totally unlike any other in the animal kingdom.

b

△ Demonstrating its unique fishing action, the **paraso** wades in the open shallow waters and holds its wings out in a huge fan in front, casting a large shadow over the water. This has two results. First, the fish actively seek out this welcoming shade to escape from the glare of the sun and second, the paraso can see and catch the fish without the distracting reflections on the surface of the water.

Mixed woodlands – swamp

GLUB

Lutasaurus anacrusus

A very specialized plant life grows in the waters of the mangrove swamp. It must be able to withstand inundation by muddy fresh water at times of flood, and by salt water at high tides. It must also be able to grow in the shifting muds brought down by the rivers. The mangroves can do this, securing their roots in the mudbanks to anchor the sediments and build them up into permanent parts of the landscape. The roots often protrude above the level of mud and water to help the plant to breathe, there being little oxygen in the mud of the swamp.

In the water there are many other varieties of plants, and these are consumed by a number of different animals. The largest is the glub, a descendant of the adaptable hypsilophodonts. It is similar in appearance to the watergulp of the tropical river swamps in the Neotropical realm (page 68), also evolved from a hypsilophodont. The resemblance is a case of parallel evolution, in which two related creatures have developed along the same lines in response to similar environmental conditions. The evolution took place independently, on opposite sides of the world. In the glub, the process has gone even further than in the watergulp. The whole body of the glub, all 2 metres (6 ft) of it, has become a swimming organ. Sinuous, lateral movements of its body and tail, aided by the tall fin down the back, drive the animal forward through the water. It steers itself by its forelimbs. The hind limbs have totally atrophied.

▽ The **glub** is totally adapted to life in the water. Its eyes and nostrils are on the top of the head to allow it to see about on the surface while the body is submerged. The front legs have paddles and a long claw for digging up the roots of the water plants on which it feeds.

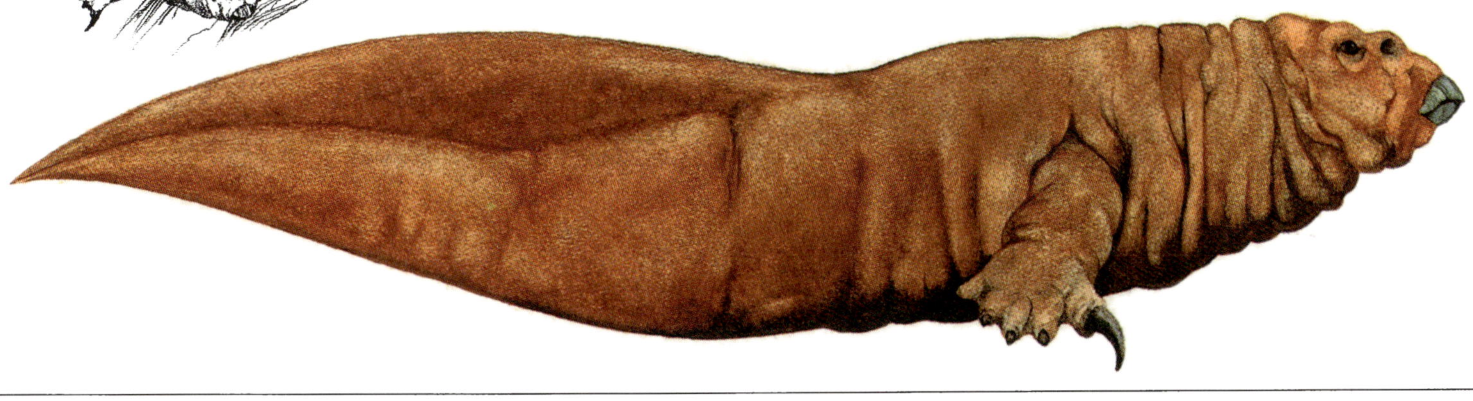

THE AUSTRALASIAN REALM

Of all the palaeogeographic realms the Australasian must be the most isolated and self-contained. The vast proportion of its bulk is a single island continent measuring about 3,500 kilometres (2,100 m) by 3,000 kilometres (1,800 m), straddling the southern tropic. Other portions of the realm consist of several large and thousands of smaller islands scattered to the north and east. Some of the eastern islands are even more isolated than the main landmass and could be regarded as small, individual zoogeographic realms of their own. The islands to the north consist of continuous archipelagoes reaching towards the great northern continent and the Oriental realm.

The history of the Australasian realm is quite simple. Up until the late Cretaceous period the continental area was part of the southern supercontinent of Gondwana. During that time rift valleys appeared and split it away from what is now the south polar continent. Ever since then it has been moving northwards, and in the past 50 million years or so it has moved from the southern polar regions into the tropical latitudes. This represents the fastest of all the continental movements since the split-up of the supercontinents, and, during this time there was no mixing of animal life from the surrounding regions. The animals present were the remains of the old Gondwana fauna - a so-called relict fauna - and these have had to change continually over the past 50 million years in order to adapt to the changing latitudes and climates. The animal life found in the Australasian realm is therefore quite unique. Over the past few million years, however, the continent has been approaching the offshore islands of the great northern landmass, and there has been some influx of animals from the north. There has also been a constant pressure from the oceanic plates to the east and the north, and as a result, island chains have formed, producing an offshore fringe of archipelagoes around the main continent. The movement has also given rise to the only major mountain chain on the continent itself, and this runs up the length of the eastern coast, from an island in the south to a tapering peninsula in the north. The remainder of the continent is a vast flat plateau, over half of which is more than 300 metres (1,000 ft) above sea level. The largest of the neighbouring islands consist of mixtures of old continental rocks - fragments of Gondwana - and new rocky material folded up from the seabed by the plate movements and forced through by new rocks emplaced as volcanoes.

The habitats of the main continent are dominated by the position of the continent across the southern desert belt of the planet. Hot air that rises and drops its rain into the rainforests at the equator spreads north and south at high altitudes and cools. At these latitudes it begins to descend and, being dry, produces arid and desert environments on any land mass that lies below. The centre of the Australasian continent is primarily desert, surrounded by dry grassland. Here live dinosaurs, relics of the old Gondwana dinosaurs, that can subsist on the poor grasses and dry scrubby bushes of the arid regions. Where an oasis of water appears, the invertebrate and fish life that thrives in it is exploited by specialized dinosaurs and pterosaurs that have evolved here and are found nowhere else in the world.

The coastal regions, particularly in the north, are more fortunate. Wet winds from the sea bring a milder and more equable climate to these areas. In the north, the forests that are produced are almost tropical jungles while, elsewhere, eucalyptus forms the characteristic woodland that clothes the eastern mountains and the coastal plains. Tree-living animals range from omnivorous arboreal dinosaurs that feed on fruits, nuts and shoots, to very specialized feeders that subsist on a diet of eucalyptus - unpalatable or poisonous to most other creatures.

The islands to the north are clothed in tropical forest, being surrounded by warm ocean. The islands to the east also have a wet climate, supporting forests of various kinds, although wide areas of grassland exist supporting a unique fauna of terrestrial grazing pterosaurs. These have developed in isolation, and they parallel the grazing pterosaurs that have evolved independently in the Ethiopian realm (pages 34–5).

The more widespread oceanic islands are not technically part of the Australasian realm. They appear during volcanic eruptions and, once cooled, they represent completely fresh habitats ripe for colonization. Once vegetation is established on an island, the animal life arrives. The islands are isolated and so they are usually inhabited by descendants of flying creatures that have crossed the oceans by air. The island habitats are fully exploited by creatures like birds or pterosaurs that have given up their powers of flight.

Tropical rainforest
Desert and desert scrub
Scrub and tall grass savanna
Short grass savanna
Temperate forest

Scrub and tall grass savanna – lakes

CRIBRUM

Cribrusarus rubicundus

The rivers that drain from the eastern mountains into the arid hinterland of the Australasian continent often seep away into the desert, or else form lakes. Lakes that form in the rainy season support an explosive burst of algae and crustaceans, and the cribrum feeds on these. Coelurosaurs have been present on the Australasian continent at least since *Kakuru* (rainbow lizard) hunted there in early Cretaceous times. It would be from a creature such as this that the cribrum evolved. In build it is rather like a conventional 2-metre-long (6 ft) coelurosaur, but the long curved jaws are armed with thousands of tiny, needle-like teeth. These strain living creatures from the fine mud and water of the deltas and lakes. An unusual feature of the cribrum is that it changes colour depending on where it is feeding. When it is feeding in the fresh water of the streams, the colour is a light grey. When it feeds on crustaceans and algae in the salty waters of the lakes, however, its skin and hair turn pink. The red coloration in the algae is concentrated in the bodies of the crustaceans that feed on it, and thus appears in the pigmentation of the cribrum that feed on them.

▽ The **cribrum** stands in the shallow water, often on one leg, and feeds by sieving the water through its finely toothed jaws. It swings its head around in a wide arc to cover as much water as possible.

△ There are few predators on the shores of the salty lakes, but when one does appear, slinking down to the water's edge to try and trap an unwary **cribrum,** the herds panic and scatter in all directions. The milling surge of pink distracts and confuses the hunter, while the voluminous curtains of spray thrown up by the dash through the shallow water conceals the direction of the fleeing herd.

Scrub and tall grass savanna – rivers

POUCH

Saccosaurus spp.

Wherever fish are abundant there will be fishing animals, including, on occasion, dinosaurs. In fact, fishing dinosaurs, such as *Baryonyx,* existed in early Cretaceous times. The rivers of the Australasian continent, however, have a unique group of fish-catching dinosaurs called the pouches. These generally belong to the same genus, *Saccosaurus,* and have evolved from the coelurosaurs. They are quite amphibious, being happier sculling about on the surface of the water and diving to the river bed than waddling about on land. Their buoyant bodies, big heads and webbed feet make the adults look very ungainly and vulnerable as they tend to their nests on river banks. The nests are built of mud and sticks, above the local flood level. The eggs, hatchlings and juveniles resemble those of a totally land-living creature, suggesting that it has not been long since the pouch evolved from a terrestrial ancestor. The swimming habits of the young pouches must be learned at their parents' side while their bodies develop the aquatic adaptations of adulthood.

In the water, the adult pouch swims gracefully on the surface, with its striped tail waving as a flag. It moves steadily with powerful strokes of its webbed hind feet and dives swiftly after fish, steering with a membrane between the forelimb and the body. The catch is held in a bag of skin beneath the lower jaw until the pouch returns to land.

△ Fish caught by the parent **pouch** are stored in the bag beneath the jaw. There they are kept safe while the dinosaur waddles up to the nest and presents them to the young offspring.

▽ Several species of **pouch** exist in the rivers of the Australasian continent. Many live together and the species are distinguished by the different patterns of colour on the tail.

▷ The webbed hind feet and steering membrane of the **pouch** enable it to swim easily through the water. Diving swiftly beneath the surface, the pouch seizes a fish between its sharp teeth, *right*.

Desert and desert scrub

GWANNA

Gryllusaurus flavus

The interior of the Australasian continent is very dry. Almost two-thirds of its area is desert or dry grassland. It is a harsh environment for living things, but not an impossible one. Many animals exist here but the only large one is the gwanna. It is the last survivor of a once widespread plant-eating dinosaur group, the iguanodonts. These dinosaurs were closely related to the hypsilophodonts, the descendants of which are now found all over the world, but the iguanodonts tended to be much larger and had more cheek teeth. On all the other continents of the world the iguanodonts were eventually replaced by the more versatile hadrosaurs, but in the Australasian continent they survived in isolation as the hadrosaurs never reached that landmass. The larger number of cheek teeth meant that the iguanodonts were in a better position than the hypsilophodonts to evolve into grass-eaters when grassy plains developed in the mid-Tertiary period. The grasses of the region are not particularly nutritious and a large animal has to range further to find enough to eat. The gwanna lives in small family groups which can move quickly from one area to another seeking fresh pastures. The physical build of the original iguanodont was quite suitable for this way of life and it has not changed dramatically. At rest, the gwanna is on all fours, with its head near the ground. When moving, it is a bipedal animal. Its longer hind legs can bear its full weight, and its body is balanced by the heavy tail as it walks or runs across the open landscape.

▽ The 3-metre (10 ft)-long **gwanna** lives in family groups of four or five adults and a number of young. The sparse grasses could not support larger herds. The gwannas' sandy colour camouflages them from a distance. They move mostly by walking or running, but when faced with sudden danger, such as finding a poisonous dingum in the grass, they leap out of the way, flashing their flank patterns as a warning to others.

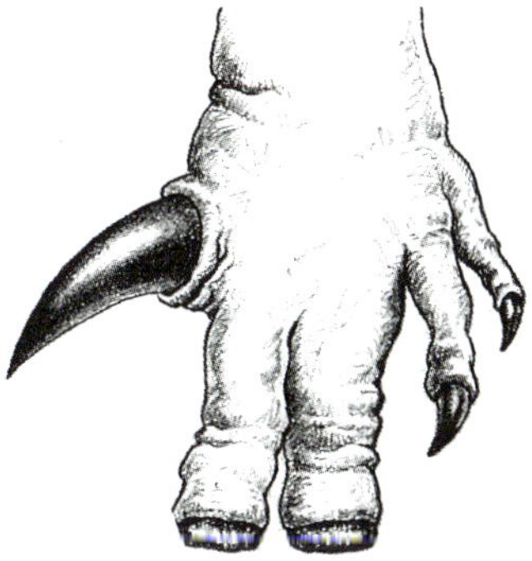

△ The **gwanna** has evolved mouth parts that can deal with tough grasses. A solid horny beak at the front of the mouth crops the grass. It is then pulled by the tongue to the cheek-pouches where it is chewed thoroughly by a series of constantly replaced grinding teeth. The brightly coloured head crest is used for signalling during the mating season. The hand *left,* can be used for walking (two hooves), for grasping (two fingers) and for fighting (thumb spike).

Desert and desert scrub

DINGUM

Velludorsum venenum

Like a big lizard, the dingum creeps through the grasses of the hot, central Australasian plains, hunting small mammals, reptiles and insects in the dry tussocks. Suddenly a dark shadow sweeps across as a hunting pterosaur hurtles down from the sky. In an immediate reaction, the dingum arches its back and dips its head. A fin of skin supported by struts of bone springs up from its curved back and presents a gaudily coloured sail to the attacker, a sudden burst of garish colour against the drab grasses. The crest on the back of the head pops up a frill of spines, each one poisonous enough to kill a large attacker. The pterosaur breaks off its attack, instinctively knowing that these colours mean danger and death, and flies away to seek easier prey.

The dingum is small for a dinosaur, measuring only about a metre (3 ft) long. It is descended from the coelurosaurs, the small meat-eaters of the Mesozoic era and, like many other Australasian animals, lives nowhere else in the world. It is a meat-eater but occasionally eats small poisonous plants. The dingum itself is immune to the poison which is then concentrated in the spines at the back of the head and used in defence. The female is quite a different animal, however, having neither poison spines nor sail. It is much larger than the male and has a more conventional coelurosaur shape. It is more secretive in its habits, avoiding predators by not being conspicuous.

▷ The **dingum's** complex mating and nesting ritual begins during the wet season. The male begins to build its nest from clay and the half-built nest is used as a display arena while it courts a female (**1**). After mating the couple continue to build (**2**). By the dry season, the nest is completed, with the female walled up inside incubating the eggs while the male hunts food (**3**). In the next wet season when the eggs have hatched, the male stands guard at a newly enlarged nest entrance while the female hunts (**4**).

1

2

3

4

◁ The bright colours of the male **dingum's** crest and sail warn would-be predators of its poisonous nature.

▷ The superficial resemblance in size and shape between the Australasian **crackbeak** and the arbrosaurs of the northern continents (pages 12–15) has come about by convergent evolution. They both evolved from lightly built running ancestors and have become lightly built tree-dwelling animals, but whereas the arbrosaurs subsist on a diet of insects and small vertebrates, the crackbeak is strictly vegetarian.

△ The **crackbeak's** bill is delicate enough to pick individual berries out of a bunch, yet powerful enough to crack open even the hardest nuts. The beak is used only for picking and cracking; the actual chewing is done by the back teeth. The horny crest on the head, along with the bright colour of the face and dewlap, are used for signalling both to mates and to enemies.

Tropical rainforest

CRACKBEAK

Fortirostrum fructiphagum

High in the luxuriant branches of the tropical forest trees in the north-east of the Australasian continent, a black and white creature throws itself from one branch to another and disappears among the foliage. At first glance it looks like an arbrosaur, one of the tree-living coelurosaurs that are found all over the world. Then it appears again, and its bright face, surmounted by a high crest, is its most obvious feature. This is a crackbeak – a tree-living descendant of the successful hypsilophodonts.

Crackbeak ancestors were lightly built running animals, evolved to run swiftly across open country. Now the legs have developed into jumping legs, and the lightweight feet have evolved for perching. The small first toe has turned round to face the rear, so the foot can now be used for grasping branches. The tail is still a long, stiff balancing organ but can now be used as a third leg; pressed against a tree-trunk it gives the animal a firm anchor while feeding. As in the arbrosaurs, the crackbeak has developed a strong system of bones and muscles in the shoulder region to help it climb trees. The front feet have become dextrous hands. Crackbeaks are found in many other parts of the world, particularly in the tropical rainforests of the Ethiopian and Oriental realms, but it is only on the Australasian continent that they are so abundant and varied.

△ The **crackbeak** hand, like those of all hypsilophodonts and most of their descendants, has a full complement of five fingers. All the fingers are long and mobile. Both the first and the fifth fingers are opposable, or able to cross over the palm of the hand in order to grasp things. These are very useful to the crackbeak while climbing in the branches and selecting food from between the twigs.

△ The **tubb's** rotund shape is unique. There is no other animal like it anywhere. It probably evolved from the hypsilophodonts that were present on the Australasian continent in the Mesozoic period. The related crackbeaks must have evolved there as well, and spread to the Oriental realm along the island chains.

Temperate forest

TUBB

Pigescandens robustus

Not all tree-living animals are active jumpers. Up in the topmost branches of the eucalyptus trees of the south-eastern and south-western corners of the Australasian continent lives a silvery blob, about 70 centimetres (2 ft) long. This is the tubb, a clumsy-looking creature moving sluggishly up and down the silvery grey branches of the eucalyptus trees, feeding on nothing but the bluish foliage. At first glance it is difficult to imagine that both the tubb and the crackbeak are evolved from the same hypsilophodont ancestors, yet there are clues. The hind feet have four toes, the first of which point backwards and allow the animal to grasp branches. The hands have five fingers, the outer two of which are opposable. The deep head has powerful jaw muscles and a sharp beak, for breaking off food from the trees.

The other physical features, however, are quite different. The body is round and not suited for rapid movement. The legs are more suited for grasping trunks than for swinging along branches. The tail is short and stubby. It is the two opposable fingers on the hand that reveal the tubb to be closely related to the crackbeak. Of the hypsilophodont descendants in other parts of the world, only the taddey of the Oriental realm (page 83) possesses this feature.

△ The **tubb** is a placid animal, moving slowly and eating nothing but eucalyptus leaves and twigs. It has neither armour nor speed for defence. Its only protection from enemies, such as birds and pterosaurs, is in its distasteful flesh. Its body tissues store up the poisons that are present in the leaves and bark that it eats.

Temperate forest – offshore islands

KLOON

Perdalus rufus

A pair of islands lies about 2,500 kilometres (1,550 m) to the south-east of the main Australasian continent. Together they are more than 1,500 kilometres (900 m) long, and so are strictly too large to be considered islands yet too small to be true continents. Geologically they contain elements of both. They consist of pieces that were once part of the main southern continent of Gondwana, yet much of their area consists of new material produced by volcanic action since Gondwana broke up.

With such a turbulent history and such an isolated position it is hardly surprising that the animal life found here is unique. Very little remains of the fauna that existed when these fragments were part of the great supercontinent. A notable exception is a group of small, primitive reptiles that lives near the north coast and has remained unchanged since Triassic times. The bulk of the animal life consists of birds and pterosaurs, many of the latter flightless. These evolved either from flightless pterosaurs that were already on the continental fragments when they broke away, and so may be related to the flightless pterosaur of the Ethiopian grasslands (pages 34–5), or from more conventional flying pterosaurs that flew to the islands and then abandoned their powers of flight.

The kloon, about 70 centimetres (2⅓ ft) long, is a typical terrestrial pterosaur of these islands. It has no wings, or indeed any trace of the forelimbs that it must have possessed at one time. It is covered in thick shaggy hair and lives secretively in the undergrowth of the forests, eating low-growing plants.

△ The **kloon,** a vegetarian, has continually growing, gnawing teeth at the front of the jaw, and broad grinding teeth at the back. This is totally unlike the tooth pattern of the ancestral pterosaur.

◁ Herds of **wandles** roam the grasslands of the southern island in leisurely moving, unconcerned herds. From an evolutionary point of view they have reached a final stage. Any influx of animals from other realms would now be a disaster for them, since they have evolved in total isolation. They are low in intelligence, are not adapted in any way for defence against predators, and could not outrun a meat-eater if they happened to meet one. Nor could they face competition from the more efficient grass-eaters that exist elsewhere.

◁ In the absence of predators, the **kloon** has adopted a totally terrestrial way of life. The forelimbs, with the wings, have disappeared completely and the hind limbs carry the weight of the whole body. The original four toes of the pterosaur foot have remained.

△ The **kloon's** feet are quite dextrous. While standing on one leg it can use the opposite foot to lift up awkwardly shaped food to the mouth. It is a slow-moving animal and spends much of its time deep in the undergrowth of the forests of the northern island.

Short grass savanna – offshore islands

WANDLE

Pervagarus altus

If there are no ground-dwelling flesh-eaters, as is the case on the islands, many flying creatures may abandon their powers of flight and take up a ground-dwelling existence. In these areas most of the habitats on the ground may be occupied by animals whose ancestors once flew. On the southern island there are large areas of grassland to the east of the central mountain chain. These are grazed by a flightless pterosaur that is related to the kloon. This is the wandle and it is quite a large animal, being about 2 metres (6 ft) high at the hips. Like the kloon, its herbivorous diet has meant that all similarity to the carnivorous flying pterosaur ancestor has been lost. Its face has taken on the appearance of one of the sprintosaurs (pages 56–7) because it must accommodate a similar tooth pattern and jaw mechanism to allow it to eat the same food. Its specializations are similar to those of the terrestrial pterosaurs, such as the flarps (page 35), of the Ethiopian realm.

Many species of wandle exist at different altitudes between the mountains and the plains, and they all eat slightly different foods – the tall grasses of the lowlands, or the shorter grasses of the foothills. One species lives even higher up and eats alpine vegetation. They are all slow-moving creatures. With no large meat-eating animals present they have not evolved defensive mechanisms such as armour, and have no need to be physically adapted for speed. The appearance of such an animal is always rather bizarre because it seems so unlike anything that exists under the normal environmental constraints that influence animal life over the rest of the world.

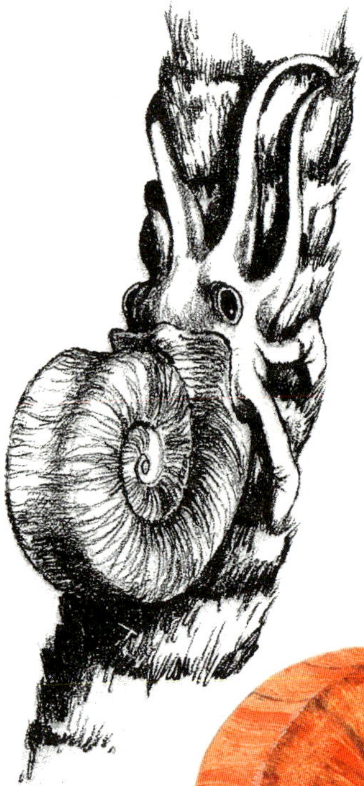

▽ The coiled shell of the **coconut grab** is flat on the bottom, providing a skid-like surface that allows it to be dragged over the sand. There are eight tentacles. The four at the rear are broad and very muscular. They are used for pulling the animal over the ground and up coconut palms, *left*. The front four tentacles are long and delicate, allowing the ammonite to reach for a coconut. The eyes can focus both submerged and out of the water. Coconut grabs usually come ashore at night when it is cooler, and dawn finds the beach criss-crossed by their distinctive trails, *right*.

▽ The **shorerunner** is a major predator of the coconut grab. Should one of the ammonites linger too long over its coconut so that the sudden tropical dawn finds it on dry land, it will be set upon by a flock of shorerunners and torn to pieces (**a**). Other prey consists of burrowing insects winkled out of tree trunks by the long jaws. The shorerunner climbs trees nimbly, balancing with its atrophied wings (**b**).

b

a

Tropical rainforest – island shoreline

COCONUT GRAB

Nuctoceras litureperus

Scattered across the vast ocean that covers almost half the globe, is a multitude of islands. These are not fragments of any continent, but have grown completely independently of Gondwana. They have appeared mainly through volcanic action, where an underwater volcano has reached the surface and cooled. The flanks of these islands are further extended by reefs built up by corals and other sea creatures. Belonging to no recognized zoogeographic realm, this array of islands is described here as part of the Australasian realm.

At the same time as the dinosaurs developed to be the most significant animals on earth, other creatures evolved to dominate the seas. A group of animals of some importance were the ammonites, cephalopods that were encased in coiled shells. The shells consisted of empty air chambers that could be used by the animal to regulate its buoyancy. The ammonites evolved into many shapes and sizes during the Mesozoic period and are commonly found as fossils in rocks that date from that time. The coconut grab is an unusual ammonite in that it can spend much of its time out of the water crawling about on land. On many of the tropical islands of the ocean it can crawl up the beach and eat coconuts, and even climb trees to find the nuts when there are none available lying in the sand or washed up on the shore.

Tropical rainforest – island shoreline

SHORERUNNER

Brevalus insularis

When new islands appear above the surface of the ocean their colonization usually follows a standard pattern. The initial living things to grow there are plants, germinating from seeds and spores borne on the wind. The first animals are always insects, again because they can be blown on the wind from distant continents. The first vertebrates to arrive are the birds and pterosaurs, winged creatures that are able to fly over the intervening ocean areas. Very often these winged creatures give up their powers of flight and take up a ground-dwelling existence, eating the plants and the insects on an island that is free from any dangerous predators.

The shorerunner is just such a pterosaur, found on one island group near the equator. It lives mainly on the beach, running here and there catching shore creatures, or pecking about in the ferny undergrowth for small reptiles and insects. It is also adept at climbing trees, using its long fingers and toes, and can run nimbly along branches using its atrophied wings for balance. The islands are no more than about five million years old, suggesting that the shorerunner's ancestors must have arrived since that time. From the flying ancestor several species have developed to populate the groups of islands, all with slightly different shapes and sizes and differing eating habits.

THE OCEANS

By far the greatest area of the planet Earth – 362 million square kilometres (140,000,000 sq m) or over 70 per is covered with ocean water. Compared with the land areas, however, the ocean is quite barren. It is largely a cold and dark place, with prolific life appearing only in the uppermost few hundred metres, where sunlight can penetrate and support a flora of drifting algae. With an average ocean depth of 3.8 kilometres (24 m), this topmost fertile layer does not represent a very large proportion of the whole. Animals also live in the chill dark ocean depths, feeding on organic debris raining down from the productive layers, but they are not numerous.

When the dinosaurs first evolved in late Triassic times all the landmasses were fused into one supercontinent called Pangaea. The rest of the planet's area was covered by one continuous ocean called Panthalassa. The division of Pangaea and the separation of all the continents has meant that Panthalassa no longer exists and the great basin that once occupied it has been shrinking ever since. Even so, its remnant – the ocean basin that stretches between the eastern end of the Palaearctic, the Oriental and the Australasian realms and the western flanks of the Nearctic and Neotropical realms still covers almost half the globe. The other ocean areas are new, having grown from rift valleys that ripped the supercontinent apart. The Tethys Sea has gone now – swallowed up by the movements and collisions of continents along the southern edge of the Palaearctic realm.

The continents are awash at the edges. The continental edges are never at the shorelines of the landmasses – they usually lie far out at sea. The resulting shallow shelf of land is called the continental shelf and may be as deep as 100 metres (333 ft). Thereafter the bottom topography drops off quickly and the seabed slopes down to the abyssal plain, the surface of the oceanic plates. Deeper portions still are found in the oceanic trenches, elongated indentations in the ocean floor where one plate is being swallowed up beneath its neighbour. Most plant and animal life is found on the continental shelves, as the seabed here is within the zone illuminated by the sun. The shelf is usually narrow along continental edges where an oceanic plate is being swallowed up, and broad where the continents are moving apart. In the late Cretaceous period, as the last parts of Pangaea broke up, the shelves were much broader than now.

Unlike the continents, the oceans have no well-defined zoogeographic realms. The great water areas are continuous with one another and have few barriers to migration.

Deep trenches
Abyssal plain and basins
Continental shelf
Land

Shorelines

SOAR

Cicollum angustalum

The waters of the world's oceans are in constant circulation, driven by the force of the winds across the surface. At certain places, particularly where the prevailing wind is blowing off the land or parallel to the coasts, the strong swirling currents bring cold nutrient-rich water up from the deep ocean floor. These areas of upwelling are found off the west coasts of the Nearctic, Neotropical and Ethiopian continents. The nutrient-rich water combined with the warm sun gives rise to blooms of floating plants in the surface waters, bringing in great shoals of feeding fish. Huge flocks of birds and pterosaurs also congregate in these areas and they hunt the fish.

One of many types of pterosaur that have evolved a fishing way of life is the soar. With a wingspan of over 4 metres (13 ft), it can hover for long periods over likely ocean areas, looking for signs of fish shoals near the surface. When a shoal is sighted, whole flocks of soars settle on the surface, and start to fish by dipping their long heads and necks quickly under the water. The animal itself is never fully submerged. After gathering a full crop it takes off again, rising clumsily into the wind from the wave tops and flying unerringly back to its nesting island. While it is hunting in flocks in the open ocean, the soar occasionally falls prey to the bird-eating plesiosaurs (page 105).

△ The **soar** nests in flocks, forming rookeries on rocky islands. Hunting trips may take several days, after which the adult brings a crop full of fish back to feed the young.

▽ The **soar** is graceful both in the sky and on the water. While swimming, it holds its wings above its back, never wetting them. Only the head and the long neck are plunged underwater.

▷ A development of the wing membrane between the hind legs and tail has produced the **plunger's** powerful swimming paddle. The swimming motion is augmented and steered by the wings, which are strong flaps of gristle. Fatty insulation is reinforced by the sleek pelt of fine hair, with its striking black and white pattern.

Shorelines

PLUNGER

Pinala fusiforme

A windswept chain of rocky islands rises above the green and white swell of the southern ocean. On the rock pinnacles, above the height at which waves surge and break, lie a number of glossy shapes, basking in the distant watery sun. These black and white creatures flop around clumsily, seeming to wobble on their bellies, pushed along by their stubby limbs, apparently quite incapable of any fast or graceful movement. Yet, when they reach a cliff edge, they rise to their hind legs and plunge head first into the foam. There they are transformed into elegant streamlined creatures, turning and darting, chasing after the shoals of fish on which they feed. In many areas the seas are so rich in fish that the fishing animals do not need to travel long distances to find them.

The plunger is a fishing pterosaur that has lost its powers of flight. Its wings are still present but modified into hydrodynamic organs that allow it a mobility in the water that its ancestors had in the air. Layers of fat have built up under the skin, and these not only insulate the animal from the chill waters but also give it the streamlined shape that allows it to swim easily. The plunger's lungs have special adaptations to withstand the great pressures found at the depths at which they hunt their food.

▽ For all its aquatic adaptations, the **plunger** still needs to come ashore to breed. It nests in vast rookeries on isolated rocks, but only in areas where the nearby ocean currents ensure a constant supply of fish.

▷ The huge pliosaurs developed specializations that enabled different species to exploit different foods. The **whulk** consumes the ocean's plankton. Its teeth have become thin, fine and tightly packed, forming a sieve structure along the jaws. A voluminous pouch has developed beneath the lower jaw. The pouch is filled as the whulk opens its vast mouth (**a**). With the mouth closed the pouch collapses, forcing the water out between the teeth and straining the plankton from it (**b**).

Temperate sea

WHULK

Insulasaurus oceanus

In the Mesozoic era the reptiles established an early mastery of the sea. Among them were the ichthyosaurs, which were the most well-adapted sea reptiles of all, with their fish-shaped bodies and tails. These no longer exist, having become extinct and been replaced by the lizard-like mosasaurs during the Cretaceous period. The turtles, slow-moving shelled herbivores, were also successful adopting a marine existence. The other main line of marine reptiles was the plesiosaurs. These quickly evolved into two main lines: the short-necked types, or pliosaurs; and the long-necked types, or elasmosaurs. Both types still exist in the modern oceans. The largest marine reptile alive today is the whulk, a pliosaur.

At 20 metres (67 ft) long, the whulk cruises the oceans of the world, but where its ancestors fed on ammonites and squid and other cephalopods, this pliosaur consumes much smaller creatures. During the Cretaceous period the shallow seas over the continental shelves produced vast volumes of plankton, tiny animals and plants that drifted in the warm nutritious waters. The shallow seas are not so extensive nowadays, but the plankton is still there. The whulk feeds on it by swallowing great volumes of water and straining out the plankton through its thousands of tiny teeth.

Temperate sea

BIRDSNATCHER

Raperasaurus velocipinnus

Sea birds wheel over the green ocean, concentrating their attention on a shoal of fish feeding in the surface waters. One by one the birds dive, splashing into the depths and surfacing again with their catches. Suddenly a great turmoil churns up the water, and in a cloud of spray, pointed heads on long thin necks burst from the surface and shoot skywards, snapping and snatching at the wheeling flock. When they subside once more into the sea the flock of birds has scattered in panic, and many of their number have disappeared.

The necks belong to birdsnatchers, the specialized, bird-eating elasmosaurs. In appearance they are very similar to conventional elasmosaurs that have existed for the past 100 million years. The seas have remained relatively unchanged, as has the fish population, and so the elasmosaur shape has proved a successful and long-lasting one. The main adaptation evolved by the birdsnatcher is its ability to catch sea birds. It lives chiefly on fish, but now and again a school of birdsnatchers will work together to seize a flock of birds from the sky. With necks held back they approach an area where birds are fishing. Then they all break the surface at the same time and thrust their necks quickly upwards, each one grabbing a bird from the wheeling flock.

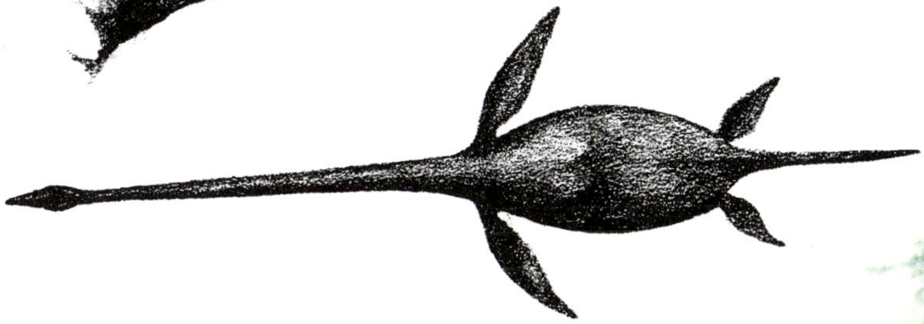

△ The conventional elasmosaur shape is evident in the outline of the **birdsnatcher.** There is a bulbous, streamlined body with paddles for limbs, and a shortish tail with a vertical diamond-shaped fin. The neck is extremely long and flexible and consists of more than 70 vertebrae.

△ The **birdsnatcher's** jaws are long and narrow, armed with pointed teeth that are angled outwards. This has been the shape of the elasmosaur head and jaws since Cretaceous times and, like the ancient elasmosaurs, the birdsnatcher lives mainly on fish. Its jaws and teeth, however, enable it to also catch birds *right*.

Open sea

PELORUS

Piscisaurus sicamalus

The pelorus is a small pliosaur and, like its relatives, it feeds on ammonites. This is the only type, however, that has the ability to attack and kill the largest ammonite of the oceans. It hunts in the warm placid waters of the doldrums, where its prey is most abundant.

There is an oily swell, and dotted here and there are the floating bulbous shapes of kraken shells, their long tentacles streaming out just below the surface, passively entrapping any small creature that happens into their entangling fibres. A splash! And a spray of water! The tranquil scene is shattered by a frenetic burst of activity around one of the bobbing shells. A pelorus has broken the surface and is wrestling with a kraken, struggling to reach its vulnerable parts. The weak tentacles loop and coil out of the water, trying to entangle the attacker, but they have little real strength. The pelorus clambers over them, wriggles through their coils, and finally plunges its sharp jaws down into the fleshy head. Again and again it stabs, until at last the tiny brain is destroyed. The writhing tentacles gradually become still, and the pelorus begins to feast, tearing off chunks of flesh as quickly as it can before the air seeps from the shell's chambers and the massive corpse sinks to the depths.

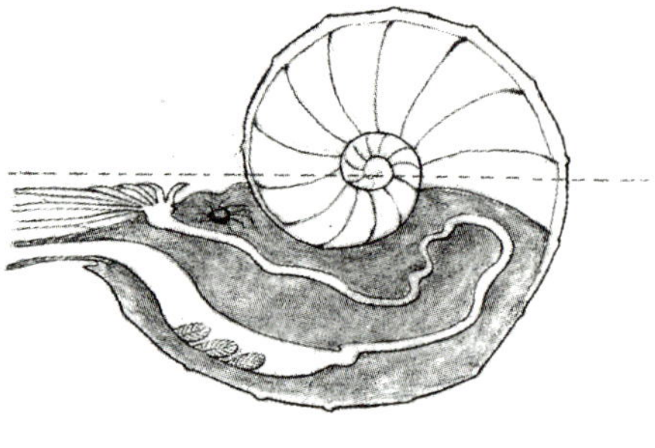

▷ The internal structure of the **kraken** is a 'scaled-up' version of the internal structure of any shelled cephalopod. The shell has a number of chambers, and the animal occupies the last one. As the kraken grows it produces more shell, moves forward and lays down a wall to create a new chamber behind. A blood vessel connecting all the chambers controls the air pressure and hence buoyancy.

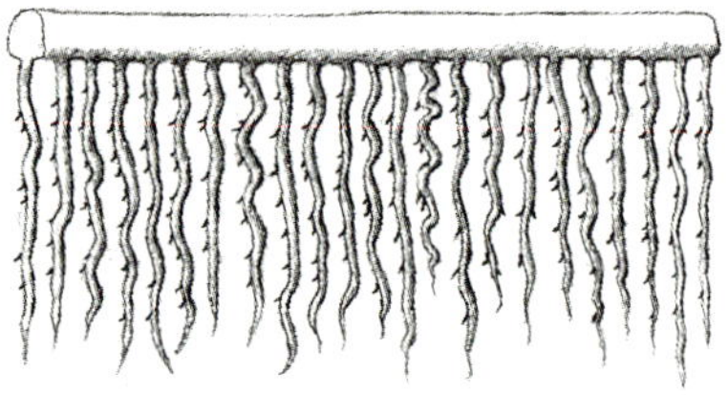

▷ The twelve tentacles with their trailing curtain of stings and hooks, radiate from the mouth at the shell's entrance.

△ The **pelorus** is immune to the stings of the kraken, but it could easily become entangled in the tentacles and drown. Its method of attack is to swim swiftly up through the curtain of fibres and clamber along the upper surface of the arms until it reaches the head. There it can stab at the soft body found in the shell chambers until the ammonite is dead.

△ The **pelorus,** at 2 metres (6 ft) long, is one of the smallest of the pliosaur group. It swims by means of its paddle-shaped limbs and the flukes on its tail. It is not a very fast swimmer, but it uses its long neck and narrow pointed jaws against its main prey, the enormous floating kraken.

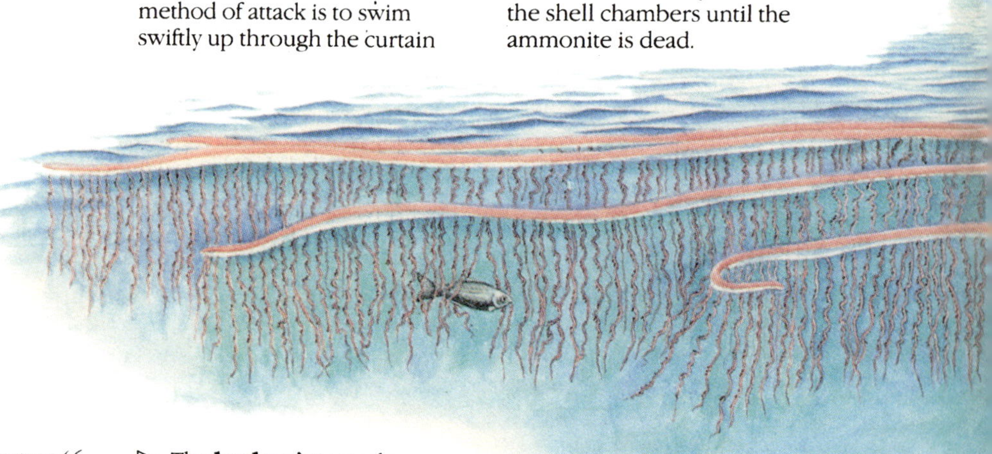

▷ The **kraken's** tentacles are used for catching animal and plant food. The trailing fibres with their hooks and stings are evolved from the suckers found on the tentacles of its ancestors. When the prey is caught it is passed to the mouth along the tentacles by muscular contraction.

Open sea

KRAKEN

Giganticeras fluitarus

The ammonites of the Mesozoic seas were mostly swimming animals that moved freely about in the ocean waters chasing small swimming creatures that they caught with their tentacles. In Cretaceous times they developed into a number of different forms. There were those with heavy shells produced in irregular coils, that spent their time crawling along the sea bed. Others were freely drifting animals, filtering microscopic food particles from the water using very fine tentacles. This second trend has reached a pinnacle with the modern kraken.

The shell of the kraken is truly enormous, some specimens reaching 4 metres (13 ft) in diameter. The shell acts as a protective armour, as well as a float to keep the animal in the fertile surface waters. It has twelve tentacles that it spreads out around itself, and each of these has thousands of trailing fibres that are armed with stings and hooks. The whole arrangement forms a deadly net that covers an area of about 20 metres (67 ft) in diameter. The kraken will eat almost anything that becomes entangled in its traps, from microscopic floating plants, to fairly large fish. Many krakens often drift in the same area of productive waters, their floating shells acting as perches for migrant birds and pterosaurs.

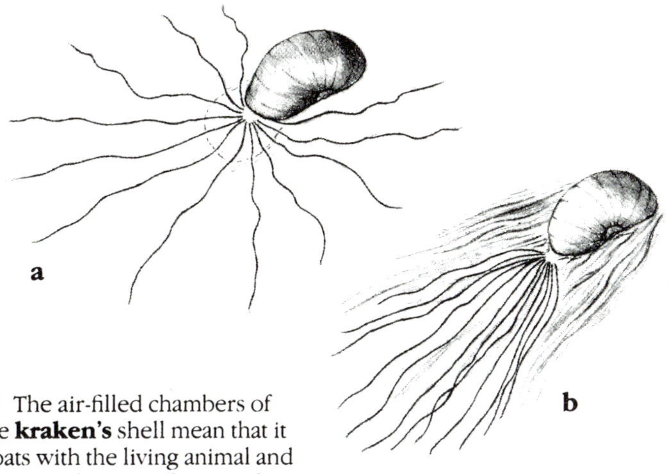

△ The air-filled chambers of the **kraken's** shell mean that it floats with the living animal and the tentacles just submerged (**a**). It moves by expelling waste water through its syphon, propelling the shell backwards and allowing the tentacles to trail behind (**b**).

CONCLUSION

The world is a complete ecosystem. The grassy plains are grazed by rajaphants and sprintosaurs. The fruits and berries in the trees of the tropical forests are eaten by crackbeaks. The twigs of the temperate and coniferous woodlands are browsed by brickets and coneaters. Mountain plants are scraped up and eaten by balaclavs and hanuhans. These plant-eating animals are, in turn, preyed upon and killed by fierce hunters like northclaws, cutlasstooths and arbrosaurs, and their dead remains are devoured by scavengers such as gourmands, and many of the pterosaurs.

Yet the numerous animals described in these pages can only provide a selective and superficial account of life on the planet. In particular, the animals are nearly all vertebrates. Many millions of other species exist, especially among the invertebrates that have been only briefly mentioned here. The grasslands, forests, deserts and mountains contain a unique fauna of tiny creatures that are just as essential to the completeness of the ecosystem as are the big spectacular beasts.

What would happen if the large reptiles were to become extinct suddenly, as has happened to the dominant forms of life many times in the past? What would develop to take the place of the large land-living vertebrates? Would it be the tiny mammals that scamper around the dinosaurs' feet, or would it perhaps be the versatile and adaptable birds, or even some novel development of the reptile stock that we just cannot imagine? As the continents are now split apart and widely separated, it is possible that the evolution would be different in each of the zoogeographic realms, thus developing a world fauna that would be more diverse than at any time in the geological past.

Whatever happens, life will survive and progress. For as long as our planet can support life, life will develop and adjust to the changing conditions.

AFTERWORD

THE SURVIVAL OF DINOSAURS IN LITERATURE

What you have been reading is a fantasy based upon a simple premise – that the Great Extinction at the end of the Cretaceous period 66 million years ago did not take place.

This is not an original premise. The remains of dinosaurs were first discovered early in the nineteenth century and the finds gripped the imagination of the public. Ever since, authors and scientists alike have been indulging in the fantasy of how things might be if these creatures still existed. The first writer of note was Charles Dickens who, in the opening paragraph of BLEAK HOUSE in 1853, described the streets of London as being so muddy that he could imagine a *Megalosaurus* waddling up them.

Lost worlds

In 1864, Jules Verne set the pattern that was to be followed by many writers. With JOURNEY TO THE CENTRE OF THE EARTH published that year, he visualized a region, in this case a system of caverns deep below the Earth's surface, in which Mesozoic animals still existed. A subterranean ocean had formed when the Earth cracked open during the 'Secondary Period' of the world (the book uses the now superseded system of geological chronology as was used in the mid-nineteenth century) and the fissures filled with sea water, and the contemporary animals that inhabited the ocean. The travellers witness a battle between a plesiosaur and an ichthyosaur.

The most noteworthy exploration of the theme was THE LOST WORLD by Sir Arthur Conan Doyle, published in 1912. In this book, a small isolated plateau in South America contained the wildlife that existed over the whole world during the Mesozoic era. The named animals sighted by the expedition include *Iguanodon* and *Stegosaurus*, and there is also a lake full of plesiosaurs and a swamp squawking with pterosaurs. The book was written during the infancy of the cinema industry, and a successful silent film was based on it in 1926. The film, using stop-motion miniature dinosaurs, emphasized the visual majesty of the creatures and initiated a vogue for dinosaurs in the visual media. The team of sculptor Marcel Delgado and animator Willis O'Brien, who were responsible for the dinosaurs in THE LOST WORLD, went on to make the most renowned 'dinosaur' film of all – KING KONG in 1933.

The 1930s and 1940s represented the heyday of the so-called 'pulp magazine'. With such eye-catching alliterative titles as STARTLING STORIES and FAMOUS FANTASTIC MYSTERIES, these magazines featured short stories and novelettes that would be classed in the then infant genre of science fiction. Many featured adventures in remote places where dinosaurs still existed. The adventure was inherently spectacular; the magazine covers usually bore illustrations of a dinosaur threatening a young woman. The dinosaur was commonly adapted from a painting by one of the famous dinosaur artists of the time – Rudolph Zallinger, whose murals were displayed in the Yale Peabody Museum, or Charles R. Knight, who had executed murals for many of the natural history museums in the United States, including the American Museum of Natural History, the Smithsonian Institution, and the Field Museum of Natural History – but with the claws and teeth absurdly exaggerated. The young woman, also with some physical features exaggerated, was usually depicted helplessly falling.

Since then, 'lost worlds' have appeared time and again in literature, in films, in comics and on television. All depictions are based on the same rationale, that a small area on the Earth's surface (or below it) has become isolated during some past period of geological time and retains the animal life existing at that time. The many locations that fiction writers have proposed for such an area include the jungles of South America, the jungles of central Africa, the Sahara Desert, an island in the Indian Ocean, an Island in the South Atlantic, an island in the Arctic Ocean, an island in the Pacific, a volcanic crater in Antarctica, a side-branch of the Grand Canyon, and remote valleys in the Rockies, the Andes and the Himalayas.

These lost worlds all seem to suffer from two rather obvious faults. The first, is that the isolation of the lost world is never absolute. Not only can the modern day explorers penetrate their mysteries, but other creatures appear to have broken in at various times. Thus, as well as dinosaurs, pterosaurs and plesiosaurs from the Mesozoic era, there are also mammoths and sabre-toothed tigers from a much later time. The originators of the genre are responsible for this fault. In JOURNEY TO THE CENTRE OF THE EARTH there are mastodons as well as plesiosaurs, and amidst the flora of the subterranean world and the famous forests of giant mushrooms, are coal forest trees from the Carboniferous period. In THE LOST WORLD there are giant Irish elk and armadillo-like glyptodonts as well as stegosaurs. The new-comers have been slipped in quite comfortably and exist in ecological balance with the animals and plants already there. In the real world such an invasion would almost inevitably have led to the extinction of the original fauna, and its subsequent replacement by the newcomers.

The second fault found in lost world stories is rather more subtle. With the exception of Tarzan creator Edgar Rice Burroughs' PELLUCIDAR series, in which the Earth is visualized as a hollow sphere with an alternative world on the inside of the shell, the lost world is always of very small area. This is necessary from a dramatic point of view, to explain the region's lack of discovery. However, it also means that the area could not possibly be large enough to sustain the huge animals that are described. If such a lost world did exist, the animals would have had to evolve specific adaptations to enable them to live in the restricted conditions. They would be quite unlike the huge and spectacular dinosaurs (usually enlarged beyond reason) that are seen to exist in the works of fiction. They would possibly have developed as did the dwarf titanosaurs and megalodontosaurs on the Indian Ocean islands in this book (pages 40–41), evolving miniature forms to cope with the diminished land area and the shortage of food. Recent palaeontology and modern zoology provides the evidence. Elephants the size of pigs developed on the islands of the Mediterranean in the late Cenozoic period, and the diminutive Shetland pony evolved its small size to survive on the sparse grazing found in the Scottish Isles. If dinosaurs had survived to this day, under whatever circumstances, they would not resemble anything like the animals we know from Mesozoic fossils. There would definitely be no *Iguanodon* or *Stegosaurus* in the 'Lost World'.

Preserved dinosaurs and others

There is another fictitious context that allows living dinosaurs to exist in modern times. The rationale is that a dinosaur is preserved in suspended animation, usually in some unlikely medium like ice or volcanic lava, and brought back to life through some even more unlikely agency such as a lightning discharge or a nuclear explosion.

Like the explanation of the lost world situation, the circumstances under which the animal is preserved merely represent a device that provides background for the story, and never pretends to be a serious investigation into the possibilities of such an occurrence. Perhaps the most influential presentation following this theme was Ray Bradbury's short story THE FOGHORN published in 1952, in which the foghorn of a lighthouse summons a dinosaur from the depths of the ocean; the sound is mistaken for a mating call. Elements of this story were incorporated into a film, subsequently released as THE BEAST FROM 20,000 FATHOMS, with the climax revealing a fictitious dinosaur on the rampage in

***Stegosaurus*, from THE LOST WORLD**
"There was a full-page picture of the most extraordinary creature that I had ever seen . . . In front was an absurd mannikin . . . who stood staring at it." The sketch of *Stegosaurus* from the diary of Maple White, the first explorer of the 'lost world', as described in the book by Sir Arthur Conan Doyle.

New York. For a low-budget film this was remarkably successful, largely due to the work of sculptor/animator Ray Harryhausen continuing the traditions established by Delgado and O'Brien in the early dinosaur dramas. The film spawned a series of popular monster films during the 1950s.

Modern versions of this theme involve the cloning of a complete dinosaur from fragments of preserved tissue. The microscopic structure of the DNA is analyzed from the cells, and induced to grow into a specimen of the entire animal. Alas, modern technology regards it as impossible to do this with fresh living tissue, let alone with something that is fossilized and 100 million years dead.

The main example of this theme, and the one that has had the most lasting effect on our culture, was, of course, Michael Crichton's 1990 novel JURASSIC PARK, and the subsequent series of films made by Stephen Spielberg and his company from 1993 onwards. As with most of Crichton's work the novel goes into a great detail about the scientific background. The premise involves cloning dinosaurs from preserved DNA found in the guts of parasitic insects fossilized in Mesozoic amber. Unfortunately for both the book and the films, dinosaur science moves on so quickly that many of the details have become very dated. The book has hypsilophodonts climbing trees (no longer believed) and the film has *Velociraptor* without the feathers we now know it to have possessed. Unfortunately, that is the fate faced by all writers who stray into the genre.

Other books, films, comics and television programmes place dinosaurs into a human context, locating them on distant planets and having them observed by visiting or shipwrecked astronauts. The rationale is that the planet is undergoing an evolution that is parallel to that on Earth but a few hundred million years behind. Again, there is no need to examine closely the science behind the premise – it is sufficient to produce the background for the story.

Alternatively, there is the fantasy scenario that portrays cavemen and dinosaurs as contemporaries. The cavemen communicate in grunts, and the dinosaurs are all ravening meat-eaters. But we are now deviating from the original premise of this book, the possible survival of dinosaurs into the present day.

It has, in fact, been argued that dinosaurs do survive today, that the birds have diverged from the coelurosaur stock in Jurassic times. In outward appearance they have changed a great deal. However, their anatomy and physiology have led some scientists to suggest that the outward differences are superficial – mere adaptations to a life of flight – and that birds are indeed close to their dinosaurian ancestors. As a result, they should be regarded as specialized dinosaurs in their own right. In this case the birds represent the surviving dinosaurs, and they have survived simply because they have changed so much from their forebears.

The question of intelligence

We cannot, it seems, live without intelligence. One important feature of the lost worlds of fiction is the presence of human beings. In both JOURNEY TO THE CENTRE OF THE EARTH and THE LOST WORLD there are examples of primitive humans living alongside the dinosaurs, presumably the result of an even later invasion than that which introduced the Irish elk and mastodons. It is unlikely that these human colonists would have remained in ecological balance with the other creatures present, when mankind's record of wildlife exploitation is considered.

The theme of intelligence in lost world works of fiction may have another source. Very often it is assumed that if the great reptiles had survived, some would have evolved a human-type intelligence and culture. Edgar Rice Burroughs introduced intelligent reptiles in AT THE EARTH'S CORE where, as the *Mahar*, they had evolved from pterosaurs, and again in TARZAN AT THE EARTH'S CORE where, as the *Horib*, they had evolved from lizards. The British Broadcasting Corporation's long running series DOCTOR WHO presented two races of intelligent reptiles, evolved from creatures that survived the extinction of the dinosaurs.

Perhaps the most prominent example in recent fiction is the animals that feature in Harry Harrison's WEST OF EDEN, and its sequels. The *Yilane* of this book have evolved from the aquatic mosasaurs. Harry Harrison's book is rather more original than the others. It is, like THE NEW DINOSAURS, based on the premise that the great extinction never took place at all. It visualizes the modern world as being populated by dinosaurs as it was in the Mesozoic era. However, on the isolated continent of North America – our Nearctic realm – the dinosaurs did become extinct, and the resultant evolution of the mammals culminated in the development of man, thus WEST OF EDEN obtains the best of both worlds.

In all these works of fiction the intelligent reptile possesses all the technical skills attributed to human beings, yet none of the finer feelings. The cold-bloodedness of the reptile is revealed through the creatures' callousness and their unemotional treatment of each other, and of any humans that stumble upon them. This depiction is usually essential to the drama of the situation.

What is often overlooked in these dramatic concepts is that

An intelligent dinosaur?
The dinosauroid, a hypothetical model of an intelligent dinosaur that may have evolved from the maniraptorans had they survived until today visualized by Canadian palaeontologist Dale Russell.

intelligence such as ours requires an endothermic, or warm-blooded physiology, to enable it to develop. Otherwise the efficiency of the brain would be extremely limited during periods of slow metabolism. This objection could easily be overcome by invoking the modern concept of warm-blooded dinosaurs, but then the dramatic effect of their cold-blooded cruelty would be lost.

The concept of the intelligent dinosaur was elaborated by the Canadian palaeontologist Dale Russell in 1981, when he published his vision of the 'dinosauroid'. Dr Russell estimated that one of the maniraptorans, *Stenonychosaurus*, was the most likely dinosaur candidate for the development of intelligence, as its brain, in relation to the size of the body, was larger than that of any known dinosaur. Furthermore, *Stenonychosaurus* was a bipedal animal and had prehensile hands with dextrous fingers. These were the very physical features that generated intelligence and civilization in the apes. Dr Russell's dinosauroid was about 1.4 metres (4½ft) tall and was very humanoid in build, with a completely upright stance, a large head, and an intelligent-looking face. Accepting his theory, by now the dinosaurs would have developed such advanced technical skills that they would have been travelling to the stars!

But is intelligence as we know it, an inevitable result of evolution? If a group of animals survives and evolves for long enough, can we assume that it will develop into a reasoning, tool-making, war-mongering, and art-appreciating civilization? Many scientists seem to think so. The project known as SETI – the Search for Extra-Terrestrial Intelligence – is founded upon such an assumption. The astronomers who listen to radio waves from the stars in order to receive and interpret intelligent signals, use what is known as the Drake equation. The equation, formulated in 1961 by astronomers Frank Drake and Carl Sagan, states mathematically that the number of civilizations that could possibly be contacted in the galaxy can be expressed by the formula:

$$N = R^*.f_p.n_e.f_1.f_i.f_c.L$$

in which N is the number of civilizations, R* is the number of stars in the galaxy, f_p is the fraction of stars with planets, n_e is the average number of Earth-like planets in a system, f_1 is the fraction of these planets on which life has evolved, f_i is the fraction of living systems in which intelligence has evolved, f_c is the fraction of intelligent beings trying to communicate, and L is the average lifetime of such a civilization. Assigning the most optimistic values to each of these factors gives scientists the possibility of contacting between 100 million and 10,000 million civilizations in the galaxy. Assigning values to these factors is largely a matter of guesswork, especially when it comes to the factor f_i, or the number of living systems that will give rise to intelligence. For the purposes of SETI, this value has been put at 1, reasoning that it is inevitable for a living system to evolve intelligence. If, however, this figure is zero, the whole equation collapses and not one extraterrestrial civilization is trying to contact Earth.

However, even on Earth, intelligence has not represented an inevitable end-product. Earth's biological systems have been successfully surviving without intelligence for 3,500 million years. Over the million years, or thereabouts, that it has been in existence, intelligence has only manifested itself as a civilization for about the last 4,000 years. Intelligence has yet to prove itself as a feature that has any evolutionary advantage at all, let alone representing the ultimate goal of evolutionary development. (The record of *Homo sapiens* as a successful long-term survivor is not good.)

Had the dinosaurs survived and continued to evolve, intelligence may indeed, have developed. However, it would not have been the kind of intuitive reasoning intelligence that we associate anthropomorphically with the term. It would be more of an animal cunning, with increasingly more sophisticated and efficient hunting techniques and cooperative abilities.

It is true to say in any case, the dinosaurs that would exist today would be quite unlike those that existed during the Mesozoic era. They would, however, be just as strange, and as magnificent, to our eyes. But, alas, our eyes would not be present to witness them.

GLOSSARY

abyssal plain The floor of the deep ocean, between the continental masses.
acid rain Rain that has a high acidity content due to dissolved gases such as sulphur dioxide.
aerodynamic Having a shape that affects a body's movement through the air.
archipelago A group of islands, usually in the form of a chain.
atrophy Of an organ, to wither away through disuse.
biped An animal that walks on two legs.
brachiate To swing along tree branches using the arms, as apes do.
browse To eat leaves. *c.f.* graze.
buoyancy The upward force exerted on a body by a fluid. An animal's buoyancy determines whether it will sink or swim.
camouflage Coloration or ornamentation of an animal that allows it to be inconspicuous against its background.
canopy The continuous cover of interlocking branches forming the roof of a tropical forest.
carnassial An arrangement of teeth evolved for shearing meat.
Cenozoic The era of geological time from 65 million years ago to the present day.
cephalopod One of a group of shellfish that have tentacles growing from around the mouth. The octopus, squid and extinct ammonites are cephalopods.
cold-blooded Lacking the ability to regulate body temperature. All modern animals are cold-blooded except for the mammals and the birds. The scientific term is poikilothermic. *c.f.* endothermic.
comet A body made of stone and ice that traces an eliptical orbit around the sun.
communal Of an animal, living as a member of a group, usually with a strict social structure. *c.f.* gregarious.
coniferous Of a tree or a forest, having cones and needle-shaped leaves.
continental drift The process by which a continent moves slowly over the surface of the Earth through geological time. This term has largely been replaced by the concept of plate tectonics, *q.v.*
continental shelf The edge of a continent that is covered by shallow sea.
convergent evolution The process whereby unrelated organisms evolve the same shape to allow them to live in similar environments. The european mole and the unrelated marsupial mole of Australia provide a modern example.
creodont A member of an extinct group of bearlike or wolf-like carnivorous mammals.
Cretaceous The final period of the Mesozoic era, 144 to 65 million years ago.
crust The outermost layer of the Earth's structure, about 40 kilometres (24 m) thick under the continents but only about 10 kilometres (6 m) thick under the oceans.
crustacean A member of a group of invertebrates with jointed legs and a hard shell. The shrimp is a modern example.
deciduous Of a tree or a forest, losing its leaves in the winter and growing new ones the following spring. The oak and ash are deciduous trees.
dextrous Able to use the hands deftly.
digitigrade Walking on the tips of the toes as opposed to the soles of the feet. Antelope and most running animals are digitigrade.
ecology The relationship of an organism to its environment and to the other organisms that live there.
ecosystem An entire community of organisms interacting with one another and their environment.
emergent A particularly large tree in the tropical forest that protrudes beyond the canopy, *q.v.*
endothermic The ability of an organism to generate heat internally by means of chemical reactions in order to regulate the body temperature.
environment The sum of the conditions surrounding an organism. Such conditions include the climate, the topography, the vegetation and the other living creatures.
equator The latitude that lies mid way between the Earth's poles. Climates are hot at the equator as the sun is nearly always overhead.
evolution The process by which animals and plants change, from generation to generation, in response to changing environmental conditions.
extinction The dying-out of all members of a particular species.
extraterrestrial Coming from somewhere beyond the Earth. Meteorites are extraterrestrial bodies.
fold mountain A mountain formed during the process of plate tectonics when one plate moves against another and crumples up the rocks and sediments in a series of folds.
fossil Any trace of an ancient animal or a plant preserved in the rocks.
Gondwana The ancient supercontinent of the southern hemisphere that has since broken up to form South America, Africa, India, Australia and Antarctica.
gizzard A cavity situated in front of the stomach in birds and some reptiles, used for breaking up the food before digestion.
granite A type of coarse crystalline rock, rich in silica. The continents, and the islands that were once part of a continent, are made up largely of granitic rock.
graze To eat grass. Special adaptations, such as teeth that resist abrasion and a sophisticated digestive system, are needed for grazing, as opposed to browsing.
greenhouse effect The raising of the temperature at the Earth's surface due to a change in the composition of the atmosphere. Some gases, such as carbon dioxide and water vapour, trap outgoing radiation like the glass of a greenhouse.
gregarious Of an animal, preferring to live in groups. This differs from communal in that it is not essential for a gregarious animal to be in a group to survive.
hibernate To avoid the harsh winter conditions by running down the body's systems and sleeping.
horn A tough substance made of keratin – the same material that forms hair – usually produced as a defensive structure.
hydrodymanic Having a shape that affects the body's movement through water.
ilium The rear bone of the hip structure, to which the backbone is attached.
indigenous Of an animal, the original inhabitant of an area.
infrared Light radiation of a frequency too long to be detected by the human eye. Heat is radiated by means of infrared waves.
insulation A substance that protects a body

from extreme heat or cold.
invertebrate An animal that has no backbone. The vast majority of living animals are invertebrates.
ischium The lower bone of the hip structure.
isotope A form of an atom that differs from other atoms of the same element, in having a different number of particles in the nucleus.
isthmus A narrow neck of land joining two broad land areas.
Jurassic The second period of the Mesozoic era, from 213 to 144 million years ago.
juvenile The young of any animal that has not reached a breeding age.
kidney An organ that filters waste products from the blood.
latitude Distance measured in degrees north and south of the equator.
Laurasia The ancient supercontinent of the northern hemisphere that has since formed North America, Europe and most of Asia.
lichen A low-growing plant that consists of an assemblage of algae and fungi cells.
mammal-like reptiles A group of reptiles that were very important in Permian times, but then declined and died out. The mammals evolved from them.
mantle The silica-rich layer that constitutes the bulk of the Earth's structure. It lies between the crust and the core.
membrane A thin sheet of living tissue.
Mesozoic The era of geological time comprising the Triassic, Jurassic and Cretaceous periods, between 248 and 65 million years ago.
metabolism The chemical reactions that take place in a living creature.
meteorite A piece of rock drifting in space or fallen to Earth from space.
migration Movement of animals from one area to another in response to changing conditions.
monsoon The type of climate found in south-east Asia in which warm wet winds are drawn in from the Indian Ocean in summer, and dry winds blow out from the interior in winter.
nectar Sugar-based substance produced in flowers to entice insects to feed, and so pollinate the flower.
oceanic ridge A volcanic ridge found throughout the oceans of the world, produced as new material rises to the Earth's surface through the process of plate tectonics.
oceanic trench A particularly deep trough on the ocean floor, usually off continental margins with extensive fold mountains or along volcanic archipelagoes. They are produced by the movements of plate tectonics and mark the plate boundaries where one plate is being drawn down and destroyed beneath another.
opposable Of a finger, able to be folded over the palm of the hand to meet the tip of another finger.
organ A structure within a living body with a particular function.
palaeogeography The study of landforms, coastlines and continental position in past geological times.
palaeontology The study of ancient animals and fossils.
Palaeozoic The era of geological time from 590 to 248 million years ago.
pampas The open grasslands of South America.
Pangaea The ancient supercontinent that comprised all the continental masses of the Earth.
Panthalassa An ancient ocean that covered all the Earth not covered by the supercontinent Pangaea.
parallel evolution The independent development of similar features in related animals through similar stages. The development of sea lions from bear-like carnivores and of seals from otter-like carnivores is an example.
parasite A creature that lives on another, gaining nourishment from it and usually harming it.
patagia Flaps of skin that form a wing in gliding animals.
peninsula An area of land surrounded on three sides by water.
Permian The final period of the Palaeozoic era, from 286 to 248 million years ago.
pheronome A chemical secreted by a creature that affects the behaviour of another.
physiology The study of the functions of an organism.
plankton Living things, plant or animal, that drift passively in the sea.
plantigrade Walking on the soles of the feet rather than the toes.
plate tectonics The large scale movement of the surface layers of the Earth. The crust and the topmost layer of the mantle form several distinct plates are continually being created along one seam – at the oceanic ridges – and being destroyed along another – at the oceanic trenches.
prairie Open grasslands in North America.
predator A meat-eating animal that actively hunts others.
prevailing wind The wind that most often blows in a particular area.
pubis The front bone of the hip structure. The shape and arrangement of the pubis is important in dinosaur classification.
scale A small plate in the skin of a fish or a reptile. It may be made of keratin – the same substance as hair – or of dentine – the same substance as teeth.
scavenger A carnivorous animal that eats meat that has already been killed.
secondary development In evolution, the re-evolution of a feature that has already been lost in a creature's ancestors.
sediment Any material, such as mud or sand, deposited by natural processes on the sea floor or the bed of a river.
sensory To do with the senses.
shelf sea A relatively shallow sea covering the continental shelf.
soar To fly at great heights using rising currents of air.
specialization A feature that develops in an organism to help it to adapt to a particular environment or way of life.
species A particular type of creature. Members of the same species can breed with one another, while members of different species cannot.
steppe Open grassland in central Asia.
stereoscopic The condition in which both eyes can be directed on to the same object and used to judge distances accurately.
stop-motion animation A technique used in the cinema, in which a model is moved into different positions and each successive position photographed on a separate frame of film. When projected, it seems that the model is moving of its own accord.
supercontinent A very large continent produced by the uniting of several smaller continents.
terrestrial Living on the ground.
Tertiary The period of geological time comprising most of the Cenozoic era, from 65 to 1.7 million years ago.
Tethys An arm of the Panthalassa ocean that once divided the supercontinents of Laurasia and Gondwana.
topography The study of the landforms of a region.
Triassic The first period of the Mesozoic era, 248 to 213 million years ago.
tropics The two lines, 23.5° north and south of the equator making the northernmost and southernmost limit of the movement of the sun during the year. The term may also refer to the area of the Earth's surface between.
tundra A region of open land in the cold regions of the Earth that is frozen in winter but thaws to a marsh in the summer.
ultra-violet Light radiation at a frequency too high to be detected by the human eye.
upwelling An oceanic current that brings cold water from the oceanic deeps to the surface.
venomous Poisonous.
vertebrate An animal with a backbone. Fish, amphibians, reptiles, birds and mammals are vertebrates.
vestigial Of an organ, present only as a trace. Humans have vestigial tails.
zoogeography The study of the distribution of animal life and the historical and ecological factors involved.

FURTHER READING

FACT
Bakker, R. T. *The Dinosaur Heresies* Longman, London, 1987
Carroll, R. *Vertebrate Palaeontology and Evolution* W. H. Freeman, Oxford, 1987
Cocks, L. R. M. (ed) *The Evolving Earth* British Museum (Natural History) and Cambridge University Press, Cambridge, 1981
Glut, D.F. *The New Dinosaur Dictionary* Citadel Press Inc, Secaucus, New Jersey, 1982
Lambert, D. *Collins Guide to Dinosaurs* Collins, London, 1983
Nicholls, P. *The Science in Science Fiction* Michael Joseph, London, 1983
Norman, D. *The Illustrated Encyclopedia of Dinosaurs* Salamander Books, London, 1985
Ridpath, I. *Life Off Earth* Granada, London, 1983
Wilford, J. N. *The Riddle of the Dinosaur* Faber, London, 1986

FICTION
Blish, J. *Night Shapes* Severn House Publishers, London, 1979
Bradbury, R. *Dinosaur Tales* Bantam, London, 1983
Burroughs, E. R. 'Pellucidar Novels': *At the Earth's Core*, *Pellucidar* and *Tanar of Pellucidar* Dover Publications, New York, 1984
Tarzan at the Earth's Core Ballantine Books, New York, 1982
Conan Doyle, A. *The Lost World* Hamlyn, London, 1986
Chrichton, M. *Jurassic Park* Knopf, 1990
Harrison, H. *West of Eden* Granada Books, London, 1984
Winter in Eden Grafton Books, London, 1986
Verne, J. *Journey to the Centre of the Earth* Penguin Books, London, 1970

The following books are not currently in print but may be found in a local library or antiquarian bookshop.
Burroughs, E. R. *The Land that Time Forgot*, *The People that Time Forgot*, *Out of Time's Abyss* McClurg Publishers, Chicago, 1924
Back to the Stone Age E. R. Burroughs Inc, Tarzana, Cal., 1937
Land of Terror E. R. Burroughs Inc, Tarzana, Cal., 1944
Savage Pellucidar Canaveral Press, New York, 1963

INDEX

N

O

P

Q

R

S

T

U

V

W

Y

Z

ACKNOWLEDGEMENTS

ILLUSTRATORS

Amanda Barlow 7, 11, 18, 20 (right), 21 (right), 22 (right), 24 (right), 25 (right), 26, 27, 39 (bottom left), 44–45, 96–97, 111 (drawn from model created by Dr. Dale Russell), 12–13 (line), 14–15 (line)
Peter Barrett (represented by Artist Partners); 72–73
John Butler (represented by Ian Fleming and Associates Ltd); 52–53, 56–57, 68–69, 70–71
Jeane Colville 32–33, 48–49, 84–85, 106–107
Anthony Duke 7, 8, 9, 10, 16, 17, 19
Andy Farmer 30–31, 42–43, 54–55, 66–67, 78–79, 88–89, 100–101
Steve Holden (represented by John Martin and Artists Ltd); 34–35, 74–75, 82–83, 90–91
Philip Hood (represented by Young Artists); 36–37, 46–47, 50–51, 58–59, 60–61, 80–81, 92–93
Martin Knowelden (represented by Virgil Pomfret); 1, 4, 28, 72–73 (black and white), 86–87, 94–95, 98–99, 104–105, 120
Sean Milne 62–63, 64–65, 102–103
Denys Ovenden 38–39, 40–41, 76–77
Joyce Tuhill (represented by Linden Artists); 20 (left), 21 (left), 22 (left), 23, 24 (left), 25 (left)

Mary Evans Picture Library, 110

Eddison Sadd acknowledgements:
Creative Director Nick Eddison
Editorial Director Ian Jackson
Editor Christine Moffat
Art Director Gill Della Casa
Indexer Mike Allaby
Production Bob Towell

Also Available from the the same author: After Man: A Zoology of the Future